THE CHALLENGE OF THE SOCIAL
AND THE PRESSURE OF PRACTICE

THE CHALLENGE OF THE SOCIAL AND THE PRESSURE OF PRACTICE

Science and Values Revisted

Edited by

MARTIN CARRIER, DON HOWARD,
and JANET KOURANY

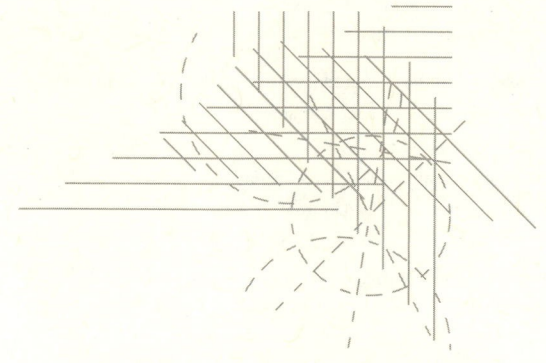

UNIVERSITY OF PITTSBURGH PRESS

Published by the University of Pittsburgh Press, Pittsburgh, PA 15260

Copyright © 2008, University of Pittsburgh Press

Manufactured in the United States of America

Printed on acid-free paper

10 9 8 7 6 5 4 3 2 1

Library of Congress Cataloging-in-Publication Data

The challenge of the social and the pressure of practice : science and values revisited / edited by Martin Carrier, Don Howard, and Janet Kourany.

    p.   cm.

  Includes bibliographical references and index.

  ISBN-13: 978-0-8229-4317-4 (cloth : alk. paper)

  ISBN-10: 0-8229-4317-4 (cloth : alk. paper)

  1. Science—Philosophy.   2. Science—Social aspects.   3. Values.   4. Science and civilization.

I. Carrier, Martin.   II. Howard, Don, professor.   III. Kourany, Janet A.

  Q175.5.C4818 2008

  306.4'5—dc22   2007041847

CONTENTS

**Part III. The Exigencies of Research Funding:
Epistemic Values and Economic Benefit**

PREFACE

The origin of the present volume was a conference on science and values held at the Center for Interdisciplinary Research (ZiF) in Bielefeld, Germany, in July 2003. The conference was organized by Don Howard (University of Notre Dame) and Martin Carrier (Bielefeld University). The organizers gratefully acknowledge the support of the ZiF, the Thyssen Foundation, and Notre Dame's Reilly Center for Science, Technology, and Values. A selection of papers from this conference were further developed to form the backbone of the present volume, and additional papers were then solicited to round out the collection. The result is interdisciplinary, diverse, provocative, and timely—a much-needed response to the challenging questions surrounding science and values.

THE CHALLENGE OF THE SOCIAL
AND THE PRESSURE OF PRACTICE

INTRODUCTION
Science and the Social

MARTIN CARRIER
Bielefeld University

Philosophical Background

SCIENCE AND VALUES ARE THOROUGHLY INTERTWINED; the two always come as a package. This is the joint message of the contributions to the present volume. As natural as this sounds today, it can by no means be taken for granted. One might be tempted to think that science simply describes what there is and that no values beyond the appreciation of knowledge are involved in the business of the scientist. Science is expected to tell us what is the case—regardless of human intentions, wishes, or fears. Social and ethical values have no place in the laboratory. In fact, however, values are involved in the process of knowledge acquisition in many ways.

There is, to begin with, the impact of science on values. Knowledge may undermine or support social and ethical values. For instance, geneticists suggest that humans, by virtue of being a comparatively young species, are more similar to one another genetically than are members of other, older species. This close genetic relation has been invoked in support of ethnic equality. Claims to biological superiority or racism are discredited by the fact that the genetic variability within ethnic groups is larger than the difference between them.

In such cases, we allow scientific knowledge to affect our commitments

1

to values. In other cases, the influence goes in the reverse direction. That is, values are used, among other things, to judge claims to knowledge. Values are brought to bear within the process of knowledge production; what qualifies as knowledge is partly determined by values. Three types of values are relevant here: ethical, cognitive (or epistemic), and social values.

Ethical values concern demands of persons for health, liberty, or integrity. They are commonly employed for judging the legitimacy of the means by which knowledge is gained, and they serve in particular to apply moral constraints to experimental setups used in research. For instance, experiments must not violate human rights. The limiting function of ethical values on the procedures adopted in science is uncontested. The experiments of Nazi doctors, for example, are universally rejected with moral contempt. Still, in some cases, as for example in stem cell experiments, it is under debate what precisely our shared ethical commitments demand or prohibit. The ethics of scientific inquiry is a vast field that has been dealt with in a large number of publications. We will focus here on the other two clusters of values in science, which have drawn far less attention.

It was widely accepted for a long period of time that science proceeds under the exclusive rule of logic and experience. In science, decisions are made based on the facts and nothing but the facts. Yet in the course of the last few decades, the philosophy of science has increasingly recognized the impact of cognitive values. Two ideas about theory choice or hypothesis assessment have been particularly influential in this respect.

The first is "Duhem-Quine underdetermination." Willard V. O. Quine famously claimed from the 1950s onward that any given set of data can always be represented by different, conceptually incompatible theories. As a result, he argued, we can get away with arbitrary hypotheses in the face of any evidence. Quine stressed that a number of alternative accounts might look utterly implausible or sound outright nonsensical. "Just about any hypothesis, after all, can be held unrefuted no matter what, by making enough adjustments in other beliefs—though sometimes doing so requires madness" (Quine and Ullian 1978, 79). Accordingly, Quine did not want to recommend the general adoption of nonstandard views in science. Instead he aimed to point to the limits of logic and experience in singling out hypotheses or theories and to illuminate the role of nonempirical values such as "conservatism" (that is, the fit of a hypothesis into the accepted body of knowledge), simplicity, or generality (Quine and Ullian 1978, 66–74).

The truth and bearing of Quine's thesis is highly contentious (Laudan and Leplin 1991; Hoefer and Rosenberg 1994; see Norton in this volume), but the feeling is widely shared that in-principle underdetermination hardly ever plays a role in actual theory choice. Such decisions are at most subject to sporadic and temporary underdetermination. But empirical equivalence of this restricted or imperfect variety is a significant factor in the appraisal of hypotheses or theories. Repeatedly in the course of history, pairs of theories emerge that entail the same consequences with regard to the available data. For instance, Henri Poincaré's 1905 version of classical electrodynamics (based on Hendrik A. Lorentz's 1904 account) and Albert Einstein's 1905 special theory of relativity were equivalent with respect to all mechanical and electrodynamic observations available until the Kennedy-Thorndike experiment was first performed in 1932. Likewise, standard nonrelativistic quantum mechanics and David Bohm's hidden variable approach, first enunciated in 1952, are indistinguishable with respect to the extant evidence (Cushing 1998, 331–35). The point here is that theories are accepted or rejected under such circumstances. In other words, the scientific community makes a choice between empirically equivalent alternatives. Special relativity was unanimously preferred to classical electrodynamics by 1920, and Bohm's account is generally discounted in favor of standard quantum mechanics. Once in a while, theories are taken to be superior even though they lack an evidential edge on their rivals.

In examples of this sort, the influence of nonempirical values on theory selection becomes conspicuous. Duhem-Quine underdetermination serves to bring to light assessment procedures that operate more covertly in the more common cases of empirical divergence. Chief among such nonempirical merits are cognitive values, that is, virtues of a theory other than adequacy to the facts, which promote the epistemic aims of science even if they are not demonstrably conducive to truth. The values mentioned by Quine—conservatism, simplicity, generality—are of this sort. So are the values provided by Thomas Kuhn, who highlights consistency, broad scope, simplicity, and fruitfulness, in addition to accuracy (Kuhn 1977, 321–22). Another list of such cognitive virtues is advanced by Peter Kosso, who features, among other virtues, "entrenchment" (i.e., fit with background knowledge), "explanatory cooperation," and testability (Kosso 1992, 36–40).

Taking cognitive values into consideration in addition to empirical adequacy fails to make theory choice unambiguous, however. It is one of Kuhn's central claims that the shared commitment to virtues like fruitfulness or generality still leaves room for picking different theories. This "Kuhn under-determination" arises from the multiplicity of cognitive values and their lack of precision. Multiplicity may lead to conflict. In a particular case of theory choice, one candidate may be more fruitful in generating new questions, whereas the alternative may possess a broader domain of application. Consequently, when values are invoked to direct a particular choice of theory, they need to be weighted and rendered precise. As Kuhn argues, subtle questions of this sort cannot be resolved by applying general rules. Rather, they need to be addressed individually using judgment and discretion. Ernan McMullin points out that this Kuhnian approach involves two kinds of value judgment. The first is "evaluation," which means that scientists assess the relevant characteristics (like simplicity or fruitfulness) of a given theory, and may do so differently; the second is "valuing," and involves the attachment of weights to such characteristics (e.g., fruitfulness is considered more important than simplicity, or vice versa). For Kuhn and McMullin, it is the room left for divergent conclusions that distinguishes value judgments from rule-governed inferences (Kuhn 1977, 330–31; McMullin 1983, 8–17).

Kuhn cites the competition between Ptolemaic astronomy and Copernican heliocentrism around 1550 as a case in point. As he argues, both accounts roughly coincided with respect to accuracy and scope. But geocentric astronomy outperformed its rival as regards its coherence with other views accepted in the science of the period. The notion that the earth remains at rest at the center of the universe fits excellently with Aristotelian physics. By contrast, Copernican theory scored better regarding simplicity or explanatory power. Whereas the core principles of Copernican theory were sufficient for explaining the bounded elongation of the interior planets and the retrograde motion of the superior planets in all their detail, Ptolemaic astronomy needed additional, tailor-made hypotheses for every single aspect of these phenomena (Kuhn 1977, 322–24; Carrier 2001, 81–92; Carrier 2008).

It deserves emphasis that Kuhn underdetermination is in no way restricted to empirically equivalent theories. It typically emerges with regard to pairs of theories that exhibit their strengths and liabilities in different areas and in different respects. Kuhn's example is the chemical revolution: The oxygen theory fared better with the explanation of weight ratios in chemical reactions, while the phlogiston theory was superior in accounting for the properties of chemical substances. In particular, only the phlogiston theory could accommodate the regularity that the properties of metals were more similar than the properties of the ores from which they were formed (Kuhn 1977, 323). The phlogistic explanation was that metallic properties were due to the presence of phlogiston as a common ingredient, whereas in the oxygen theory metals were assumed to be elementary. Balancing such assets and deficiencies defies an algorithmic procedure and can only be accomplished by invoking an elaborate system of values.[1]

However, it is suggested by Kuhn's account that cognitive values fail to finish the job of hypothesis appraisal or theory choice. Cognitive values need adjustment and fine-tuning in order to generate unambiguous assessments, and it is plausible that noncognitive considerations at least sometimes supplement methodological reasoning. The room left for theory choice by the facts and the cognitive commitments shared within the relevant scientific community is sometimes filled by the appeal to social values. Such values express claims to participation or protection of social groups; they grant or deny social groups influence on political processes or defense against material or social disadvantages. Social values may express the interests of a social group or class, economic aspirations, or commitments to social equality.

Sociologists of science have shown in what measure social values influence the assessment procedures in the sciences. Social values are usually not enforced by external political pressure. Cases of this sort occur, to be sure, but they are rare and of limited impact. Galileo's fight for heliocentrism against the church or Lyssenko's assault on Darwinian evolution through variation and selection, strongly backed by the authorities of Soviet Russia, remained short-lived episodes that exited quickly from the scientific scene. Yet, typically, social values function differently; they exert their influence without the use of coercion. The impact of social values is not due to outside pressure; it rather emerges from within the scientific community, namely, when strong social attitudes encounter diffuse evidence.

Take the example of nineteenth-century brain research. The rise of physiology increasingly supported the conviction that the structure of the brain should be able to reveal psychological features. Physiological parameters considered relevant during this period were brain weight, asymmetry between hemispheres, amount of convolutions, and distinctiveness of the frontal lobe. Judged by present knowledge, these quantities are uninformative; psychological differences between humans do not depend on raw parameters of this kind. However, the history of nineteenth-century brain research is replete with success stories claiming that correlations between such parameters and psychological or social properties had been established. The striking feature is that these alleged findings unfailingly reproduced the prejudices of the period. For instance, men, and mathematicians in particular, were purported to possess richer brain convolutions than women. Another ostensible finding was that physiological differences in the brain indicated descent (i.e., allowed a distinction between European and non-European origin), social rank, intelligence, or personality (Hagner 1999, 251–60).

This example reveals that attitudes regarding social groups may create expectations that are imposed, as it were, on unclear data and eventually dominate the interpretation of the experience. However, even in the case at issue the malleability of the data was limited. The observed results were sometimes resistant to particular interpretations and undermined the associated psychophysiological hypotheses. For instance, the initial assumption of a correlation between intelligence and brain weight was abandoned in the late nineteenth century when it was discovered that the brain weight of the then famous Göttingen mineralogist Friedrich Hausmann ranked in the lower third of the usual range (Hagner 1999, 259). While the facts remain recalcitrant to some degree, it is undeniable that social values have a considerable impact on scientific thought (Carrier 2006, 165–67).

Of course, the influence of social values might be restricted to the incipient or immature phases of theory growth. McMullin surmises that the noncognitive factors that contribute to the adoption of a theory are gradually sifted out as the theory matures. If a theory continues to thrive, it is eventually backed by a purely epistemic track record, and the fact that social values remain part of the confirmation basis in the long run weigh against the theory (McMullin 1982, 23).

However, philosophy of science is rife with dissenters to this view, among them Philip Kitcher, who argues that the goal of science cannot adequately be delineated by reference to epistemic values alone. What we want from science is "significant truth," and what "significance" amounts to is inevitably tied up with pragmatic criteria and social values. Significance is what matters to us, which is why social values are not driven out of maturing theories but are intrinsic to the practices of science. The question arises as to which values these should be. The appeal to "our" values and purposes often hides neglect of particular groups of people. The interests of some groups are sometimes disregarded, and scientific results are sometimes produced that violate specific interests. Indeed, recognition of certain truths may have a detrimental impact on some people. The result is that the pursuit of scientific research is in need of a justification that takes into account the well-considered interests and judgments of everyone concerned. Kitcher demands, in short, that the research agenda should be shaped by social values (Kitcher 2001, 65, 71, 118–19; Kitcher 2002, 552–56; Kitcher 2004, 53).

This preliminary exposition suggests that science and values, cognitive and noncognitive values alike, are intimately interwoven at various levels. As a result, the role of values is unsuitable for distinguishing between "sound science," which is allegedly value-free, and "junk science," which is contaminated by nonepistemic values. Rather, noncognitive value commitments on the part of the scientists enter into science of either type (Brown 2001, 198–200). Against this background, it is an important challenge to analyze the nature and bearing of value commitments in science and to distinguish beneficial from detrimental consequences of such commitments. In particular, adopting certain values might support or undermine the objectivity and credibility of science, it might promote or interfere with the advancement of understanding nature and the betterment of the human condition.

The Structure of the Book

This brief overview of the philosophical debate on the role of values in science and their impact on scientific inquiry shows that values have moved from the fringes to center stage.

Section I on "The Play of Values within the Core Area of Scientific Research" is opened by John D. Norton. He dissents from the background views just outlined and refuses to accord values a central role in science. He argues that the thesis that evidence necessarily fails to determine theory is, at best, speculation. It is supported only by a simple hypothetico-deductive account of confirmation and not by its more sophisticated variants or by other accounts of inductive inference. He also argues that the instances of empirically equivalent theories routinely displayed in the underdetermination literature are self-defeating. The very fact that empirical equivalence is readily demonstrable means that we cannot rule out the possibility that the two theories are merely notational variants of the same theory. As a result, he concludes that the literature on the underdetermination thesis has failed to demonstrate a gap between evidence and theory, which values could be called upon to close.

Margaret Morrison acknowledges the impact of social values on theory development and theory choice but attempts to disentangle the social from the cognitive. She addresses the use of reductionist methods in the development of population genetics from the 1910s onward and explores the pertinent debate as an example of the intersection between cognitive and social values. She shows how this reductionist approach, pursued in particular by Karl Pearson and Ronald Fisher, changed the focus of certain aspects of evolutionary theory away from biological individuals to statistical averages and model populations. The proponents of this approach were strongly committed to eugenics and used reductionist strategies to support racist goals. Morrison identifies concerns regarding the latter social values as a source of criticism of reduction and mathematical abstraction. Her plea is to separate scientific hypotheses and their cognitive evaluation from social concerns. As she argues, scientific approaches should not be assessed by the detrimental ends to which they might be put.

Helen Longino places an even stronger emphasis on values by highlighting their importance in the context both of scientific pursuit and scientific justification. First, she questions the distinction between cognitive and social values. She draws attention to alternative sets of cognitive values and makes clear how these values do different political work. Second (and relatedly), in agreement with what is now called social epistemology, she questions the traditional contrast between social values and scientific objectivity. The traditional view entails that social values distort science—that is, drive science off its epistemic course. Social epistemologists claim, by contrast, that some social procedures and social values constitute the basis of objectivity. For example, feminist and other kinds of political values and preferences broaden the substantive range from which theories emerge and the perspectives from which they are judged. Indeed, scientific objectivity demands inclusion of and tests across a wide spectrum of viewpoints, including value-based viewpoints. Objective knowledge is generated by a social procedure that is committed to maximum inclusion.

Janet Kourany goes one step further. Longino's pluralism welcomes all social values into science; in this sense it continues to adhere to the ideal of objectivity as neutrality. The confrontation of contrasting social values in a regulated procedure is assumed to produce the sort of objective knowledge that the old ideal of value-free science was unable to deliver. Kourany proposes a different conception of scientific objectivity, the "ideal of socially responsible science," which sanctions the inclusion of only certain social values in science, namely, those that answer to the needs of a just society. Socially responsible science favors research projects that motivate social reform and help to bring it about. Feminist research programs are only one kind of example. The goal is at least to realize both the social and epistemic objectives historically associated with the old ideal of value-free science. The goal, in short, is science with a human face.

In concluding this section, Jay Rosenberg insists on the separation between values internal to science or "constitutive values," on the one hand, and external, social values or "collateral values," on the other. As he claims, science essentially aims at an explanatory representation of the world; internal values are consequently cognitive or epistemic values. Improving the human condition and changing social practices may be appropriate aims connected with science, but they are not values constitutive of science per se.

Section II on "The Demands of Society on Science: Socially Robust Knowledge and Expertise" is devoted to the role of science in political decision-making. Expert advice is highly influential in many areas of politics; expertise is sometimes viewed as hooking science up with the political sphere. By contrast, Peter Weingart stresses that the approaches of science and politics continue to diverge. The two spheres cannot be intermingled without hurting one of them. One attempt to connect the two is participation, which, however, tends to compromise the reliability and legitimating power of expert knowledge. Another option is letting experts make decisions but at the expense of a glaring conflict with the democratic value of representation. As a result, seeking truth and exerting power remain contrastive activities. This means that the notion of "socially robust knowledge," which is intended to include elements from both spheres, is insufficient as a conceptual tool for controlling science and technology in democratic societies.

Roger Strand attempts a more affirmative interpretation of the notion of "social robustness." Experts are often faced with concrete problems, such as dredging a harbor and removing toxic chemicals, which cannot be solved adequately on a purely scientific basis. The actual systems experts need to attend to are highly complex, and their precise makeup is frequently unknown. Under such conditions, expert knowledge is often of high precision, to be sure, but of uncertain relevance since it usually refers to simplified situations. The in vitro knowledge experts tend to draw on is of unknown accuracy and reliability for solving concrete problems of policy advice. As a result, expert advice is unable to remove the risks accompanying technological challenges. Strand concludes that science and expertise are of limited help in many areas of policy making.

Socially robust knowledge draws on additional nonscientific sources in order to produce in vivo knowledge that is sometimes more helpful in coping with the concrete challenges we face.

Christopher Hamlin likewise addresses the role and nature of expertise, its reliability in matters of public importance, and the options for feeding knowledge into the political process. An expert is not adequately conceived as a specialist scientist. Rather, expertise needs to be conceptualized in a way different from science. For instance, whereas precision is important in science, it does not play a crucial role in many practical challenges. More often, conclusions pointing in the right direction will do. Further, expertise does not aim at comprehensive principles but at specific solutions to particular problems that do justice to the messy details of the subject matter, respect the values of the clients and stakeholders, and can be realized given the channels and constraints of policy making. Hamlin outlines a history and philosophy of expertise that places doing rather than knowing at center stage and regards trust more than reason, prudence more than proof, multiple options more than universals as the pivot of expertise.

Section III on "The Exigencies of Research Funding: Epistemic Values and Economic Benefit" addresses the impact of economic pressure on the epistemic credentials of science. Applied science, not pure research, is the focus of interest—and funding. Science is expected to be instrumental in solving practical questions, driving technology ahead, and promoting economic growth. This primacy of application puts science under pressure to quickly supply solutions to concrete challenges. There is a dominance of technological interests that threatens to narrow the agenda of research and encourage sloppy quality judgments.

James R. Brown points out that, at least with respect to pharmaceutical research, quality judgments are not only sloppy but tendentious. They comply with the economic interests of the corporations that underwrite the corresponding studies. In testing the comparative efficacy of new drugs, it matters who sponsors the tests. In addition, there is a striking imbalance in research topics favoring expensive and patentable solutions to health problems. Regarding the research agenda, new drugs are always given preference over diet and exercise. As Brown argues, the prevalence of economic interests creates an epistemic predicament in medical research that can best be resolved by increasing public funding and eliminating patents altogether.

Martin Carrier agrees that economic forces may do harm to the epistemic credentials of science. Three potentially threatening features of commercialized research are identified, namely, a biased research agenda, secrecy over research outcomes, and a neglect of epistemic aspirations. However, there are counteracting factors or influences within applied research that limit the detrimental impact of these features. One of the relevant arguments is that secrecy often backfires. Research sites operating behind closed doors are cut off from the benefits of cooperation. This is why a policy of open labs is much more widespread in industrial research than was anticipated. Given this exchange of research outcomes in commercialized science, disinterestedness of the researchers is not necessary (and might even be impossible to achieve). Rather, what is important is control of judgments and interests by contrasting judgments and interests. Competition among companies could contribute to stimulating such reciprocal control and might provide resources for a pluralist conception of rationality that is advocated in parts of social epistemology (see, for instance, Longino's chapter in this volume).

. Matthias Adam inquires into the actual role and the legitimate status of disinterestedness as a moral and an epistemic norm. His focus is likewise on pharmaceutical research. Adam objects to Brown's recommendation of a socialization of medical research: public funding, he emphasizes, is no royal road to disinterestedness. Rather, strong and partial interests of the remaining actors of the health systems can be expected to shape the research agenda under such conditions, and by inverting the commercial reward system, academic achievement rather than practical medical success would probably determine a scientist's rank and reputation. In response to Carrier's idea that competition between partial interests might neutralize the negative epistemic impact of such interests, Adam points out that such a neutralization is only accomplished if the participants adhere to some extent to the ideal of objectivity. It follows that pluralism is not sufficient for generating objectivity; it needs to be supplemented with some form of individual commitment to disinterestedness.

Conclusion

In the last quarter century, new epistemic and institutional conditions have been imposed on scientific research. Science has benefited from financial support to an unprecedented degree, but this huge influx of money was not motivated by the desire to learn more about supersymmetry or dark energy. It was rather based on the expectations that scientific research was a driving force of technical innovation and, consequently, of economic growth, and that scientific research was useful to provide advice on policy. The first expectation was primarily directed at the natural sciences, and the second focused on the social sciences such as economics. As a result, we have witnessed a thorough commercialization and politicization of science, as science has been increasingly subordinated to practical goals. Thus, the question naturally arises of what the dominance of the applied perspective does to science and, in particular, whether it hurts the epistemic aspirations traditionally tied to scientific knowledge. Conversely, we observe a growing impact of the outcomes of scientific research, or what passes as such, on ethical commitments and social organization. Cases such as genetic engineering or the preference for deregulated markets (as is urged by economic theory) demonstrate the far-reaching influence of science on society. As a result, the converse question is what the rapid growth of practically relevant scientific knowledge does to society.

In the present volume, these questions are addressed from contrasting and interdisciplinary perspectives. The divergence of the positions on the actual or proper role of values in science offers wide room for orientation or disagreement. The broad spectrum of issues addressed is emphasized by the tridisciplinary composition of the book. The collection includes chapters with a marked historical accent, studies that stress sociological considerations, and articles that pursue a philosophical approach. Philosophers, sociologists, and historians of science attempt to explore the different aspects of the complex interrelations between science and society and to analyze the changes they have undergone in the past decades. While it is true that science has been connected with technical and political ends since the time of Bacon, it is only in the course of the twentieth century that science was able to live up to these expectations. The complex fabric produced by interweaving cognitive and social threads merits the sort of interdisciplinary exploration to which the present volume is intended to contribute.

NOTE

1. Whereas this Kuhnian train of thought introduces values by contrasting them with rules, more rule-governed approaches to theory choice like Lakatos's can also be regarded as value-laden (Carrier 2002).

REFERENCES

Brown, James R. 2001. *Who Rules in Science?*. Cambridge, MA: Harvard University Press.

Carrier, Martin. 2001. *Nikolaus Kopernikus*. Munich: Beck.

———. 2002. "Explaining Scientific Progress: Lakatos's Methodological Account of Kuhnian Patterns of Theory Change." In *Appraising Lakatos: Mathematics, Methodology, and the Man*, ed. George Kampis, Ladislav Kvasz, and Michael Stöltzner, 53–71. *Vienna Circle Library*. Dordrecht: Kluwer.

———. 2006. *Wissenschaftstheorie: Zur Einführung*. Hamburg: Junius.

———. 2008. "The Aim and Structure of Methodological Theory." In *Rethinking Scientific Change and Theory Comparison: Stabilities, Ruptures, Incommensurabilities?* ed. Léna Soler, Howard Sankey, and Paul Hoyningen-Huene. Dordrecht: Springer.

Cushing, James T. 1998. *Philosophical Concepts in Physics: The Historical Relation between Philosophy and Scientific Theories*. Cambridge: Cambridge University Press.

Hagner, Michael. 1999. "Kluge Köpfe und geniale Gehirne: Zur Anthropologie des Wissenschaftlers im 19. Jahrhundert." Reprinted in *Ansichten der Wissenschaftsgeschichte*, ed. Michael Hagner, 227–68. Frankfurt: Fischer, 2001.

Hoefer, Carl, and Alexander Rosenberg. 1994. "Empirical Equivalence, Underdetermination, and Systems of the World." *Philosophy of Science* 61:592–607.

Kitcher, Philip. 2001. *Science, Truth, Democracy*. Oxford: Oxford University Press.

———. 2002. "The Third Way: Reflections on Helen Longino's *The Fate of Knowledge*." *Philosophy of Science* 69:549–59.

———. 2004. "On the Autonomy of the Sciences." *Philosophy Today* 48 (Supplement 2004): 51–57.

Kosso, Peter. 1992. *Reading the Book of Nature: An Introduction to the Philosophy of Science*. Cambridge: Cambridge University Press.

Kuhn, Thomas S. 1977. "Objectivity, Value Judgment, and Theory Choice." In *The Essential Tension: Selected Studies in Scientific Tradition and Change*, 320–39. Chicago: University of Chicago Press.

Laudan, Larry, and Jarrett Leplin. 1991. "Empirical Equivalence and Underdetermination." *Journal of Philosophy* 88:449–72.

McMullin, Ernan. 1983. "Values in Science." In *PSA 1982 II. Proceedings of the 1982 Biennial Meeting of the Philosophy of Science Association: Symposia*, ed. Peter Asquith and Thomas Nickles, 3–28. East Lansing, MI: Philosophy of Science Association.

Quine, Willard V. O., and J. S. Ullian. 1978. *The Web of Belief*. 2nd ed. New York: Random House.

PART I

THE PLAY OF VALUES WITHIN THE CORE AREAS OF SCIENTIFIC RESEARCH

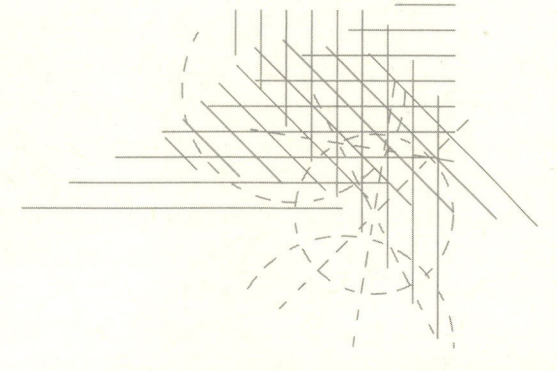

❖ 1

MUST EVIDENCE UNDERDETERMINE THEORY?

JOHN D. NORTON
University of Pittsburgh

A CCORDING TO THE UNDERDETERMINATION THESIS, all evidence necessarily underdetermines any scientific theory. Thus it is often argued that our agreement on the content of mature scientific theories must be due to social and other factors. In this chapter, I will draw on a long-standing tradition of criticism to argue that the underdetermination thesis is little more than speculation based on an impoverished account of induction. I will argue that a more careful look at accounts of induction does not support an assured underdetermination or the holism usually associated with it. Finally, I will urge that the display of observationally equivalent theories is a self-defeating strategy for supporting the underdetermination thesis. The very fact that observational equivalence can be demonstrated by arguments brief enough to be included in a journal article means that we cannot preclude the possibility that the theories are merely variant formulations of the same theory.

Some will be quite comfortable with these claims. Others will not. Recalling that science is a human activity, they will expect that, one way or another, social, cultural, political, ideological, and a myriad of other human engagements can become enmeshed in the final product of science. These factors may be peripheral in that they affect only the external conditions under which science is practiced or scientific questions will be pursued. Or they may become involved in the very content of science itself.¹ In the latter case, the real questions are not whether this happens, but when does it happen: how much do these factors determine the content of scientific theories; and how does it come about that they do? Given the very great complexity of scientific practice and the premium scientists themselves place on eradicating intrusion of external factors into the content of their theories, it would seem safest to answer these questions by direct study on a case-by-case basis. However, many scholars in science studies have proven unable to resist an easy answer to the question that assures a place for these factors in all mature sciences in advance of any direct study.² This easy answer will be the subject of my chapter, and I will call it "the gap argument":

> Step 1. No body of data or evidence, no matter how extensive, can determine the content of a scientific theory (underdetermination thesis). But there is universal agreement on the content of mature scientific theories. Therefore, there is a gap: at least a portion of the agreement cannot be explained by the import of evidence.

> Step 2. My favorite social, cultural, political, ideological, or other factor is able to account for what fills the gap. Therefore, my favorite factor accounts for a portion of the content of our mature scientific theories.

The second step of the argument has obvious problems. It is hard enough in particular cases to establish that these social or other factors played a role in deciding the content of a theory. If it happens, the scientists themselves are eager to hide it, so the point must be made in contradiction with the scientist's overt claims. Thus one might despair of finding a properly grounded way to enlarge the claim so that it covers all of science at once. Or at least it is daunting, if the claim is to say anything more than the banality that something other than data, we know not what, seems to be involved.

This second, dubious step of the argument is not my concern in this chapter. My concern is the first step. The underdetermination thesis has long been a truism in science studies, accepted and asserted with the same freedom that philosophers now routinely remark on the impotence of logic to provide us with a finite axiomatization of arithmetic. Yet the underdetermination thesis enjoys no such status in philosophy of science, where it is hotly debated. If you consulted a recently written companion to philosophy of science you would find the following synopsis: The thesis "is *at the very best* a highly speculative, unsubstantiated conjecture. Even if the thesis can be expressed intelligibly in an interesting form, there are no good reasons for thinking that it is true" (Newton-Smith 2001, 553, emphasis in original). My goal will be to display why I think this is a fair assessment. I do not intend to give a survey of the now expansive literature on the underdetermination thesis.[3] Rather, I plan to assemble two strands of criticism from it that I find especially cogent and put them in a strong and simple form.[4]

In the following section, I will state what I take the underdetermination thesis to assert and also what I believe it does not assert. I will also summarize three strategies that have been used to support the thesis: those based in Duhemian holism, Quinean holism and an argument based on the possibility of displaying pairs of observationally equivalent theories. This last argument is inductive, since the pairs displayed are supposed to illustrate in examples what is claimed to happen universally. In the subsequent section, I will lay out the first principal objection to the thesis: it depends essentially on an impoverished and oversimplified account of the nature of inductive inference. A brief survey of major accounts of inductive inference will show that they support virtually none of the properties of inductive inference presumed in the context of the underdetermination thesis. For example, in most accounts of induction, if two theories are observationally equivalent, their common observational consequences might still supply differing evidential support for the two theories. Then I will address the displaying of observationally equivalent theories as a support of the underdetermination thesis. I will suggest that such displays are self-defeating. If it is possible to establish that two theories are observationally equivalent in argumentation compact enough to figure in a journal article, they will be sufficiently close in theoretical structure that we cannot preclude the possibility that they are merely variant formulations of the same theory. The argument will depend upon a notion of physically superfluous structure. Many observationally equivalent theories differ on additional structures that plausibly represents nothing physical. I will also introduce the converse notion of gratuitous impoverishment; some artificial examples of observationally equivalent theories are generated by improperly depriving structures in one theory of physical significance.

The Underdetermination Thesis

The underdetermination thesis asserts that no body of data or evidence or observation can determine a scientific theory or hypothesis within a theory. This simple assertion requires clarification in two aspects. First, if the thesis is to be relevant to the use of data or evidence in science, the data must bear on theory by way of induction or confirmation relations, for this is how data or evidence bear on theory in science. Second, underdetermination is intended to convey the notion that the adoption of a particular theory cannot be based solely on evidential considerations. This may happen merely because the theory in question is one for which no strong evidence may be found. The underdetermination thesis is intended to have universal scope, so it must also apply to cases in which there does appear to be strong evidence for the theory. In both cases, underdetermination is explicated as the assured possibility of rival theories that are at least as well confirmed as the original theory by all possible data or evidence. The rival theories cannot be better confirmed, or else they would become our preferred choice. So the underdetermination of theory by data or evidence means that rival theories exist that are equally well confirmed by all possible evidence or data.

What the Thesis Does Not Assert

(Merely) De Facto Underdetermination The thesis does not merely assert that the data or evidence actually at hand *happens* to leave the theory underdetermined. That certainly is possible, especially when new theories are emerging. In other cases, sometimes real, sometimes contrived, it may be very hard to procure the requisite evidence. The underdetermination thesis is much stronger. It asserts that in all cases, no matter how long and ingeniously evidence collection may proceed, underdetermination will persist.

(Merely) Sporadic Underdetermination The thesis also does not merely assert that cases can arise, either contrived or natural, in which some part of a theory transcends evidential determination. Such cases clearly can arise. Lorentz insisted that there was an ether state of rest, even though Einstein's work in special relativity made it clear that no observation could determine which it was, of all the candidate inertial states of motion. The underdetermination thesis is much stronger; it asserts that all theories are beset with this problem.

Humean Underdetermination The thesis is also distinct from the most famous of all philosophical problems, Hume's problem of induction, or simply *the* problem of induction. This is the purported impossibility of providing a noncircular justification of induction. No matter how often we have seen that bread nourishes and fire burns, we cannot infer that they will continue to do so. We beg the question, Hume told us, if we ground our conclusion in the assertion that patterns in the past have continued into the future, for that assertion itself presumes the tenability of induction on patterns. If one accepts Hume's skepticism, no evidence from the past on bread or fire will determine their behavior in the future through justifiable inference. This form of underdetermination has been called "Humean underdetermination" (Laudan 1990, 322–24). It is distinct from the underdetermination of the underdetermination thesis since it denies the possibility of induction outright, as opposed to addressing a failure of the determining power of induction. This form of the thesis trivializes underdetermination. If one denies that induction is possible, a fortiori one must deny it any interesting properties, such as a power of unique determination.[5]

Underdetermination by Grue In Goodman's (1983) celebrated example, our past observations of green emeralds confirm the hypothesis "All emeralds are green" and also equally the hypothesis "All emeralds are grue," where "grue" means "green if examined prior to some future time T and blue otherwise." The problem of grue can be given a narrow or a broad reading. Neither coincides with the underdetermination thesis. In the narrow reading, grue reveals an underdetermination in the import of evidence within a particular class of syntactic theories of confirmation: those that allow an A that is a B to confirm that all As are B, where we are free to substitute anything grammatically valid for A and B. This is a narrow problem for a particular class of theories, solved by various restrictions on the confirmation theory. Quine (1970b), for example, solved it by restricting A and B to natural kind terms. My own feeling is that the problem requires a deeper reappraisal of the nature of induction (Norton 2003). In the broader reading, the problem of "grue" is just the problem that any pattern can be projected in arbitrarily many ways, with none supposedly distinguished by the earlier history. This broader reading of the problem of "grue" just makes it a version of Humean underdetermination, which I have argued above is distinct from the underdetermination thesis.

Arguments for the Thesis

I will develop three arguments that have been used to justify the underdetermination thesis: local, global, and inductive. Before turning to them, I should mention one basis for the thesis that withstands little scrutiny: the thesis has become such a commonplace that it has entered that elusive body of knowledge labeled "what everyone knows." That is, it is justified by the fallacy, *argumentum ad populum*.[6]

Duhem's Local Justification An important justification for the underdetermination thesis is derived from Pierre Duhem's celebrated analysis of the logic of hypothesis testing in science (1954, chap. 6). Duhem remarked that a hypothesis cannot be tested in isolation. A hypothesis must be conjoined with others before observationally testable conclusions follow from it. Consider one of Duhem's examples (§6.2), the hypothesis that the vibration in a ray of polarized light is parallel to the plane of polarization. If an observable consequence is to be deduced from it, the hypothesis must be conjoined with many other hypotheses in geometric and wave optics, and even with the fact that the darkening of a photographic plate measures the intensity of light. But what if the observable consequence fails to obtain? We might discard the original hypothesis or we might protect it from refutation by discarding one of the other hypotheses. So Duhem was led to ask "Are certain postulates of physical theory incapable of being refuted?" (§6.8). The gap between Duhem's analysis and the underdetermination thesis is easy to close. If we can accommodate our theory to recalcitrant observations in many ways, then we have many distinct systems of hypotheses that accommodate the observations. Thus—and this is the step I shall return to below—the observations are powerless to decide between the systems. It is not clear whether Duhem himself would have wanted to follow the modern literature in taking this last step. He completed his discussion by describing how the good sense of physicists eventually would lead to a strong decision in favor of one of the competing theories (§6.10).

Quine's Global Justification The underdetermination thesis entered the modern literature in the conclusion of Quine's "Two Dogmas of Empiricism" as a quiet by-product of Quine's project of dismantling the venerable analytic-synthetic distinction and of the notion that meaningful statements are reducible by logical construction to those about immediate experience. In the resulting account, experience was allowed only to place looser constraints on our conceptual system. Quine's view[7] was most typically and most fully described in suggestive but elusive metaphors, such as this celebrated passage from "Two Dogmas" (1951, §6).

> The totality of our so-called knowledge or beliefs, from the most casual matters of geography and history to the profoundest laws of atomic physics or even of pure

mathematics and logic, is a man-made fabric which impinges on experience only at the edges. Or, to change the figure, total science is like a field of force whose boundary conditions are experience. A conflict with experience at the periphery occasions readjustments in the interior of the field. Truth values have to be redistributed over some of our statements. Reevaluation of some statements entails reevaluation of others, because of their logical interconnections—the logical laws being in turn simply certain further statements of the system, certain further elements of the field. Having reevaluated one statement we must reevaluate some others, which may be statements logically connected with the first or may be the statements of logical connections themselves. But the total field is so underdetermined by its boundary conditions, experience, that there is much latitude of choice as to what statements to reevaluate in the light of any single contrary experience. No particular experiences are linked with any particular statements in the interior of the field, except indirectly through considerations of equilibrium affecting the field as a whole.

Duhem and Quine's Justifications Compared There are obvious similarities between the Duhemian and Quinean paths to underdetermination. Both depend on the idea of some logical distance between hypothesis and experience that leaves the latter unable uniquely to fix the former. There are also significant differences that I have tried to capture with the labels "local" and "global." In the local approach, we generate observationally equivalent sets of hypotheses by adjusting individual hypotheses in some given system. The resulting equivalent systems will differ at most locally, that is, in just the hypotheses that were altered. In the Quinean approach, however, one posits ab initio that our total system of knowledge, including science, ordinary knowledge, and even mathematics and logic, is not fixed by experience, which can only impinge on it from the periphery.

When the full weight of this difference is felt, one sees that Duhem and Quine really have very different conceptions. Duhem's approach is narrowly focused on the confirmation of scientific hypotheses by scientists in actual scientific practice. The underdetermination Quine envisages permeates our entire conceptual system, extending to physical objects, Homer's gods, subatomic entities, and the abstract entities of mathematics (1951, §6). These radically different, alternative total systems envisaged by Quine are not the sort that could be generated by multiple applications of Duhemian adjustments. Indeed Quine finds such an extension implausible (1975, 313–15; see also Hoefer and Rosenberg 1994).

Exactly because Quine's underdetermination extends all the way through to the abstract entities of mathematics themselves, it is not clear that his version of underdetermination is properly presented as a limitation of the reach of evidence in the context of the establishment of scientific theories. Is the relation over which the underdetermination prevails the relation of confirmation of inductive inference? This relation is not usually invoked in fixing the abstract entities of mathematics. Once we fix the abstract entities of our conceptual system, might Quine's underdetermination no longer affect the determining power of evidence in a scientific theory, if ever it did? Quine's statements about underdetermination are so brief and metaphorical as to preclude answers and, in my view, even a decision as to the precise nature of his notion of underdetermination and whether it is well supported.

Intermediate positions are available. Longino (2002, 126–27; 1990, 40–48) argues that data has no evidential import, in the absence of background assumptions. While the context invoked is not as broad as that invoked by Quine, she has in mind something grander in scope than merely the other scientific hypotheses Duhem envisaged; the background assumptions may include substantive methodological claims, such as the assumption that correlations are attributable to common causes. That evidence does not uniquely determine theory is in turn traced to the lack of unique determination of these background assumptions.

Inductive Justification: Instances of Observationally Equivalent Theories

The underdetermination thesis amounts to the claim that for any theory, there is always a rival that is equally well supported by all possible evidence. The credibility of the thesis is often supported inductively by displaying such pairs of theories. They are usually labeled "empirically equivalent theories" or "observationally equivalent theories," and this is usually taken to mean, especially in the latter case, that the two theories have identical sets of observational consequences. Observational equivalence would appear to be a necessary condition for theories to be empirically equivalent; any difference in observational consequences could direct us to the possible evidence that would distinguish the theories. (I will urge below that it is not sufficient.)

This justification of the underdetermination thesis is inductive. We see a few examples of how a given theory may be paired with an observationally equivalent rival. We are to conclude that this will always be possible, which result is just the underdetermination thesis, as long as we judge that evidence cannot decide between observationally equivalent rivals.

There are not many examples. I find it helpful to group them with labels inspired by pearls: natural, cultured, and artificial.

Natural Pairs   These are pairs that have arisen in the real development of science. The best-known examples are these. In Newton's original version of mechanics, there was an absolute state of rest which transcended observational specification. We form an observationally equivalent pair by taking versions of Newtonian mechanics with different absolute states of rest. Einstein's special theory of relativity and a suitably adjusted version of Lorentz's ether theory agree on all that can be observed about the slowing of moving clocks and shrinking of moving rods, yet they disagree on the theoretical account behind it.[8] In the 1920s, Cartan and Friedrichs showed how to construct a space-time theory that returned all the same motions as Newton's theory of gravitation. Yet their theory did not represent gravitation as a field, but as a curvature of space-time, modeled after Einstein's general theory of relativity. Finally it is a standard part of the lore of quantum mechanics, that two distinct quantum theories emerged in the 1920s, the matrix mechanics of Heisenberg, Born, and Jordan, and the wave mechanics of de Broglie and Schrödinger. They were soon shown to be equivalent.[9]

Another pair that is often mentioned in this context is the nonrelativistic quantum mechanics of particles and Bohm's (1952) hidden variable theory, although they are not strictly observationally equivalent. The former theory assigns no definite position to a quantum particle in most of its states; the position is brought to be by the act of measurement. The latter theory always assigns a position to the particle and it just becomes manifested on measurement. In both cases there is a probability distribution associated with the resulting positions and they will agree in all ordinary circumstances.[10]

Cultured Pairs   These are instances of credible scientific theories contrived by philosophers specifically for philosophical ends. The best-known examples come from geometry. The geometry of a surface can be recovered from the behavior of measuring rods transported over its surface. Following an earlier, related example of Poincaré, Reichenbach described how one could conform almost any desired geometry to a given surface. If the behavior of transported rods did not conform to the desired geometry, one posited the existence of a universal force field whose sole effect was to distort the measuring rods away from the true lengths of the geometry to those that were observed. The same observables would be compatible with many suitably adjusted pairs of geometries/universal force fields. Glymour (1977) and Malament (1977) have described observationally indistinguishable space-times. An observer can receive light signals only from that portion of space-time encompassed by the observer's past light cone. Since space-times can be conceived in which this observable portion is only a small part of the total space-time, even for observations made over an infinite lifetime, a full specification of that observable portion fails to fix the entire space-time. Newton-Smith (2001, 535–36) reports another intriguing example. We routinely represent the continua of space and time by the real continuum. No measurement of finite precision can distinguish that continuum from one merely consisting of rational numbers.

Artificial Pairs These are instances of incredible scientific or quasi-scientific theories contrived by philosophers specifically for philosophical ends. In van Fraassen's constructive empiricism, we are enjoined to accept the observational consequences of a theory but not to assent to the truth of its theoretical claims. Kukla (1998, chap. 5) employs a variant of this view to generate empirically equivalent but logically incompatible pairs of theories. We start with any theory T and construct T' as that theory which asserts the truth of the observational consequences of T but the falsity of all of its theoretical claims (where van Fraassen merely withheld judgment). Kukla's proposal is immediately identifiable as the latest in a venerable lineage of tortured narratives that portray a world in which we would be mistaken to believe that things are really just the way they seem. Descartes wrote of a deceiving demon purposefully planting false beliefs. More recently we hear of the world created a few thousand years ago, complete with its ancient fossil record; or of ourselves created five minutes ago with a lifetime of memories in place; or the nightmarish fable of the evil scientist who has placed our brains in vats of nutrients where they are fed perfectly realistic sensations of a world that is not there. All these deliver revisionist theories, logically distinct from the standard theories from which they stem, but with identical observational consequences.

The Underdetermination Thesis Neglects the Literature in Induction and Confirmation

An Impoverished Hypothetico-Deductive View of Confirmation

This is probably the most common of all objections leveled against the underdetermination thesis and it has been raised in many guises. The concern is quite straightforward and quite fundamental. Since the underdetermination thesis makes a claim about inductive or confirmatory relations, one would expect the thesis to be supported by the very long tradition of work in induction and confirmation. Yet the expositions of the underdetermination thesis seem to make only superficial contact with this literature and the arguments for the thesis, notably the three sketched above, seem to depend entirely on a flawed account of the nature of induction. That is, they invoke an impoverished version of hypothetico-deductive confirmation, through which

> Evidence E confirms theory T, if T, possibly with auxiliaries, logically entails E.

> If two theories T and T' are each able logically to entail E, then T and T' are confirmed *equally* by E.

One sees immediately that this account of confirmation is troubled by an excessive permissiveness. The standard illustration of the concern is frivolous conjunction. If some theory T entails evidence E, then so does the strengthened theory $T' = T \& X$, where X is any hypothesis, no matter how odd or frivolous. $T \& X$ then enjoys the same level of evidential support as T.

It is exactly this permissiveness that renders impoverished hypothetico-deductivism an uninteresting option in scientific practice and in the induction literature. Yet it is exactly this permissiveness that the arguments for the underdetermination thesis seek to exploit. For this impoverished version of hypothetico-deductive confirmation appears to be the relation routinely invoked in accounts of the underdetermination thesis. Empirically equivalent pairs of theories, in this literature, typically turn out to be pairs of theories that have the same observational consequences. They are automatically read as being equally confirmed by these consequences. Or we are to generate a rival theory by making compensating adjustments to the hypotheses of the original theory. The rival is equally confirmed by the common observations only if all that matters in the confirmation relation is that the two theories have identical observational consequences. In his discussion of empirical underdetermination in "On Empirically Equivalent Systems of the World," Quine (1975, 313) wrote: "The hypotheses [that scientists invent] are related to observation by a kind of one-way implication; namely, the events we observe are what a belief in the hypotheses would have led us to expect. These observable consequences of the hypotheses do not, conversely, imply the hypotheses. Surely there are alternative hypothetical substructures that would surface in the same observable ways." And surely we are intended to conclude that these alternative hypothetical substructures are worthy rivals to the originals.

What We Learn From the Literature in Induction and Confirmation

There are many accounts of induction and confirmation, and I will say a little more about them below. First, however, I want to collect a number of generalizations about what is to be found in that literature:

> Most accounts of induction and confirmation do not admit any simple argument that assures us that evidence must underdetermine theory.

The impoverished hypothetico-deductive confirmation relation is exceptional in admitting such an argument.[11] Most accounts do not give us the opposing result, however. Excepting cases in which a fairly rich framework is presumed, most accounts of induction do not assure us that evidence can determine theory. As a matter of principle, most accounts of confirmation leave open whether evidence determines or underdetermines theory.

> Most accounts of confirmation do not allow that theories with identical observational consequences are equally confirmed by those consequences.

One might wonder how we could ever have taken seriously an account of confirmation that tells us otherwise. Consider a revisionist geology in which the world is supposed created in exactly 4004 BC complete with its fossil record. Standard and revisionist geologies have the same observational consequences, but we surely do not think that the fossil record confirms a creation in 4004 BC just as strongly as the ancient earth of standard geology. In standard geology, detailed analysis points to precise datings for the standard geological eras. What in the fossil record points exactly to 4004 BC as the date of creation of revisionist geology and not, say, 8008 BC or last Tuesday? Indeed the fossil record would seem to strongly disconfirm the 4004 BC creation.

> Many accounts of confirmation do not restrict evidence to deductive consequences of hypotheses or theories.

As Laudan and Leplin (1991) illustrate, there are many cases of evidence that is not a logical consequence and logical consequences that are not confirmatory. The latter arises when both a hypothesis and all credible competitors entail the same consequence. Laudan and Leplin's example is the hypothesis H that regular reading of scripture induces puberty in young males and the consequence

E of the onset of puberty in some particular young males who read scripture (1991, 465).

> Many accounts of confirmation allow evidence to bear directly on individual hypotheses within a theory rather than merely supporting the entire theory holistically.

This is not to dispute Duhem's remark that hypotheses often need the support of auxiliary hypotheses if observational consequences are to be derived from them. It does dispute the idea that the obtaining of these consequences bears inductively solely on the conjunction of all the hypotheses. Consider the observation of spectral lines in sunlight corresponding to the helium spectrum. The inference from the presence of helium in the sun to the observed spectrum requires numerous additional assumptions about the optics of cameras and spectrographs. But the observed helium spectrum is strong evidence for the presence of helium in the sun and at best tangential evidence for the optical theory of the camera.

Three Families of Theories of Inductive Inference

The general conclusions above about accounts of induction derive from the existing literature of criticism of the underdetermination thesis and from a synoptic survey that I have developed of theories of induction (Norton 2005, 2003, §3). Below I sketch some general themes from the survey relevant to the underdetermination thesis. Accounts of induction can be grouped into three families, each driven by a basic principle with the variant forms of the theories generated by the need to remedy weaknesses in the principle. They are inductive generalization, hypothetical induction, and probabilistic induction.

Inductive Generalization The first family is based on the principle that an instance confirms its generalization. The simplest and best-known form is just enumerative induction: that some *A*s are *B* confirms all *A*s are *B*. The weakness of this account is the very limited reach of the evidence. Accounts that have sought to extend its reach include Hempel's satisfaction criterion of confirmation, which expands the logic used from syllogistic to first-order predicate logic, Mill's eliminative methods, and Glymour's bootstrap. The last two seek to extend the reach of inductive generalization by allowing insertion of new terms in the induction relation. Mill's methods allow us to reinterpret conditions of necessity and sufficiency as causal relations; Glymour's bootstrap allows us to employ hypotheses from the theory under test to convert evidence in an observation language into instances of hypotheses in the language of theory. These accounts are struggling with a restriction antithetical to underdetermination. The import of the evidence in this family is neither vague nor diffuse; it is precise and narrow—so narrow that much effort is expended in seeking to widen it. This narrowness of inductive generalization has allowed Glymour to emphasize that his version of the relation both contradicts holism, in that it allows evidence to support particular hypotheses within a theory, and contradicts the underdetermination thesis in that the relation allows evidence to accord different strengths of support to observationally equivalent theories (1980, chaps. 5, 9).

In inductive generalization, an instance confirms the generalization. It will often be the case that the generalization entails the instance; but there is no principle *simpliciter* that anything that entails the instance is confirmed by it. One readily finds cases in the family in which evidence confirms hypotheses that do not entail the evidence. Laudan and Leplin (1991, 461) report the simplest case, known since antiquity as "example": that this crow is black confirms that an as yet unexamined crow is also black. (Laudan and Leplin proceed to display many more examples of cases of evidence that are not consequences.) As a result, this family of accounts of confirmation tends to be immune to arguments for underdetermination that depend upon theories being confirmed by their deductive consequences.

Glymour's bootstrap allows us to use auxiliary hypotheses to infer from evidence to instances of the hypothesis to be confirmed. In an extension of this idea, the inference from evidence to hypothesis is rendered fully deductive by using suitably strong auxiliary hypotheses. The resulting approach goes under many names, including "demonstrative induction," "eliminative induction," and "Newtonian deduction from the phenomena." What makes the approach nontrivial is that we may find that we have already strong, independent evidence for the needed auxiliary hypotheses, so that the inference from evidence to hypothesis can be made deductively without taking any further inductive risk. Most striking is that, through this scheme, the evidence determines the hypothesis supported; it is deduced from it. I have described how this approach has been used in a number of cases. For example, it was used shortly after 1910 to force quantum discontinuity on the basis of the evidence of observed thermal radiation spectra in a climate in which researchers were disinclined to accept the result (Norton 1993, 2000).

Hypothetical Induction The second family of accounts of induction is based on the idea that a hypothesis is confirmed by its deductive consequences. This idea in its rawest form is just the impoverished account of hypothetico-deductive confirmation that lies behind the underdetermination thesis. Exactly because the idea is so weak as to admit rampant underdetermination, the account is never actually used in its raw form. It is always used in a form that tames its indiscriminateness by the addition of further requirements. These additional requirements generate the various accounts of induction in the family.

For example, for evidence E to confirm H, H must not only entail E but, in what I call "exclusionary accounts," it must also be shown that if H were false E would very likely not have obtained. Satisfying this additional condition automatically excludes the rivals envisaged in the underdetermination thesis. The additional condition proves fairly easy to secure. Mayo (1996) has described how it arises in the context of traditional statistical hypothesis testing and has shown how it can be extended to other cases through her notion of a severe test. Exclusionary accounts supply numerous examples that contradict the underdetermination thesis. In a controlled study, test and control groups are randomized so that the only systematic difference between them is that the test group only is administered the treatment. It then follows that any subsequent difference between the groups is evidence just for the hypothesis of the efficacy of the treatment and does not confirm any other rival hypothesis that proponents of the underdetermination thesis might envisage. There are other more indirect ways of securing the exclusionary requirement that if H were false then E would very likely not obtain. One of the most elegant arises in the "common cause" or "common origin" arguments (Salmon 1984, chap. 8; Janssen 2002).

Other natural augmentations of the impoverished hypothetico-deductive confirmation relation have the same effect of deflecting underdetermination. One popular approach is to demand that the hypothesis being confirmed is simple, with complicated rivals thereby precluded.[12] In another, "inference to the best explanation," the hypothesis supported by the evidence must also be the best explanation of the evidence (Lipton 1991). In another that I have called "reliabilism" (Norton 2003, §3), the hypothesis confirmed must in addition be produced by a method known to be reliable. The demand underpins the dismissal of what appear to be empirically adequate hypotheses as ad hoc. The complaint is that they were not generated by a reliable method, but by artful contrivance and are thus not candidates for evidential support.

Probabilistic Accounts The third family of theories of induction, probabilistic accounts, is based on the idea that scientists carry degrees of belief that are governed by a calculus and that the import of evidence propagates through the belief system by the rules of a calculus. The inspiration for the view came from the theory of stochastic processes—originally games of chance—for which the probability theory was created. The probability calculus remains the favored calculus and the dominant approach is called "Bayesian" after the theorem used centrally in propagating belief.

These probabilistic accounts are the least hospitable to holism and the underdetermination thesis. In them, the import of any item of evidence E throughout the entire system of belief can be traced, hypothesis by hypothesis. To see its import for any hypothesis H, one merely compares the posterior probability $P(H|E)$ with the prior probability $P(H)$. The dynamics of the resulting redistributions can contradict starkly with the natural suppositions of Duhemian holism. As Salmon (1975, §4) showed, one can have the following circumstance: A hypothesis H and auxiliary A entail some observation O that fails to obtain. While the falsity of O refutes the conjunction H & A, it may confirm H and may also confirm A! (See also Howson and Urbach, chap. 4.)

Standard within the lore of Bayesianism are limit theorems that describe how the accumulation of evidence forces convergence of belief onto the correct hypothesis. The theorems work very well in the contexts in which they apply, but they do require nontrivial assumptions to delimit those contexts. For example, if we have two hypotheses with the same observational consequences, no accumulation of observations can alter the distribution of belief between them. That is set in advance by the prior probability distribution.

Probabilistic accounts also give us the most general understanding of the possibility that hypotheses may be confirmed or disconfirmed by evidence that is not a deductive consequence. It arises whenever a hypothesis H is not probabilistically independent of the evidence E. That relation is expressed formally as the inequality of $P(H|E)$ and $P(H)$, which immediately asserts that the evidence E confirms or disconfirms H.

Probabilistic accounts, in their Bayesian formulation, provide the most extensive and versatile of all accounts of inductive inference. The view they provide of inductive relations is quite antithetical to both holism and a necessity of underdetermination of theory by evidence. Even if one thinks that they give a partial account of inductive relations that works well only in some domains, they still preclude any unqualified claims of holism and underdetermination.

The Display of Observationally Equivalent Theories Is Self-Defeating

Observationally Equivalent Theories as Variants of the Same Theory

We have seen that the possibility of displaying observationally equivalent theories is taken as evidence for the underdetermination thesis. One can of course generate observationally equivalent theories on the cheap by merely permuting terms. To use Quine's (1975, 319) example, we could merely switch the words "electron" and "molecule" in modern science. It is generally agreed, as Quine asserts, that the resulting pair of theories is not distinct in a sense relevant to the underdetermination thesis. They are merely notational variants of the same theory; they are merely the same theory dressed in different clothes. Quine (1975, 322) judges this to be so also in the case of Poincaré's alternative geometries. A concern frequently expressed in the literature on the underdetermination thesis is that all the commonly displayed examples of observationally equivalent theories are defective in some way like this and that this masks some deeper failing (see, for example, Newton-Smith 2001, 532–33). If just one or two examples turned out to be suspect, that would be a curiosity. But when all do, we have a phenomenon that needs to be explained. In the following, I will try to make this concern more precise and explain why we should expect all the examples developed in the literature to be suspect. I will urge that the difficulty with this inductive justification of the underdetermination thesis runs far deeper than the obvious concern that a powerful, universal conclusion is being grounded in very few instances. I will conclude that that the display of observationally equivalent theories is a self-defeating strategy for establishing the underdetermination thesis.

Every Well-Established Example Must Be Suspect

I shall seek to establish the following thesis:

> If it is possible for us to demonstrate the observational equivalence of two theories in a tractable argument, then they must be close enough in theoretical structure that we cannot preclude the possibility that they are variant formulations of the same theory.

Two aspects of the thesis must be stressed. First, it does not apply to all observationally equivalent theories. It is specifically restricted to those whose observational equivalence can be demonstrated in the sort of compact argumentation that can appear in a paper in the philosophy of science literature. That extra restriction is essential to the argument that establishes the thesis, for the ease of demonstration of observational equivalence will be used to ground a deeper similarity. Second, the thesis does *not* assert that the relevant observationally equivalent theories *must* be variant formulations. All that is asserted is that this is a strong possibility in every case in which we can demonstrate observational equivalence. But that doubt in the special cases is sufficient to defeat the inductive justification of the underdetermination thesis. For such an induction must proceed from instances known to be observationally equivalent, that is, those whose observational equivalence can be asserted in the literature with proper warrant. This induction now turns out to be suspect, for an induction from instances all of which are suspect is itself suspect.

A General Argument

The basic idea of the argument for the above thesis is that the manipulations needed to demonstrate observational equivalence are only likely to be tractable when the two theories are very close in structure, since the theoretical structures are our only practical way of circumscribing the observational consequences of a theory. The argument is developed in a series of steps:

1. If we are to demonstrate that two theories have identical observational consequences, then we must have a tractable description of their observational consequences. The description may be oblique or indirect. But it must be possible, otherwise, we would have no way to prove something about the observational consequences.

2. The description of the observational consequences of the first theory will most likely make essential use of the central theoretical terms of the first theory. It might happen that the observational consequences of the theory could be described compactly without recourse to these theoretical terms. But then we would have a very odd theory indeed—one whose entire observational content can be compactly described without its theoretical structure. Such an odd theory would be ripe for excision of superfluous theoretical structure!

3. For the same reason, the description of the observational consequences of the second theory will most likely make essential use of the central theoretical terms of the second theory.

4. If we are to be able to demonstrate observational equivalence of the two theories, the theoretical structures of the two theories are most likely very similar. While it is possible that they are radically different, if that were the case, we would most likely be unable to demonstrate the observational equivalence of the two theories. For the theoretical structures are what systematizes the two

sets of observational consequences and a tractable demonstration of observational equivalence must proceed by showing some sort of equivalence in these systematizing structures.

5. The two sets of theoretical structures may be interconvertible without loss; or they may not be. In the latter case, there would be additional structures present in one theory but not in the other. However any such additional structure will be unnecessary for the recovery of the observational consequences. That follows since the additional structure has no correlate in the other theory, yet the other theory has identical observational consequences. Thus any additional structures will be strong candidates for being superfluous, unphysical structures.

6. We conclude that pairs of theories that can be demonstrated to be observationally equivalent are very strong candidates for being variant formulations of the same theory.

Physically Superfluous Structure

Consider once again the natural and cultured examples of observationally equivalent theories listed above. (I will return to the artificial examples below.) They are not manifestly variant formulations of the same theory. Lorentz's ether theory is physically distinct from Einstein's special relativity; Lorentz's theory posits an ether state of rest, whereas Einstein's does not. Indeed, we saw above that observationally equivalent theories are not generally equally confirmed by their observational consequences. That conclusion would be unsustainable if observationally equivalent theories were mere variant formulations, for then they must be equally confirmed or disconfirmed by all evidence.

How are we to reconcile this? The answer is that step 5 of the argument above asks us to take a particular view of certain structures in pairs of observationally equivalent theories. In some cases, it can happen that the observationally equivalent theories are intertranslatable without loss. Matrix mechanics, in suitably rich formulation, can be converted into wave mechanics; and wave mechanics can be converted into matrix mechanics, without loss of structure in either direction. Fully intertranslatable theories are automatically strong candidates for being variant formulations of the same theory. More typically, we have a more complicated situation in which there is a loss of structure. Lorentz's theory can be converted into Einstein's by the simple expedient of asserting that the results of time and space measurements with slowed, moving clocks and shrunken, moving rods are the true measures of time and space. But we must also discard the ether state of rest, since it has no counterpart in Einstein's theory. According to Einstein's theory, it is superfluous structure corresponding to nothing in the physical world.

Step 5 of the above argument enjoins us to treat all such superfluous structure as representing nothing physical. And it does so with good reason. For we have two theories with a common core fully capable of returning all observations without the additional structures in question. Moreover, in some of the examples to be visited in a moment, the additional structure may lack determinate values. For example, as stressed by Einstein famously in 1905, the ether state of rest of Lorentz's theory could be any inertial state of motion. Because of the perfectly symmetrical entry of all inertial states of the motion in the observational consequences of Lorentz's theory, no observation can give the slightest preference to one inertial state over another. So its disposition is usually understood not to be an unknowable truth but a fiction.

Observationally Equivalent Theories Must Not Be Judged Variants of the Same Theory

I want to emphasize here that nothing compels us to accept that observationally equivalent theories are merely variant formulations of the same theory. As long as the theories are logically distinct, one can always insist that differences of theoretical structure correspond to something physical, even if the differences are opaque to observation or experiential test. This situation arose with Lorentz's electrodynamics, which is observationally equivalent to special relativistic electrodynamics. Long after Einstein's celebrated work of 1905, Lorentz insisted upon the correctness of his theory and its distinctness from Einstein's special relativity: his theory incorporated an ether state of rest not to be found in special relativity. The mainstream of physics found Lorentz's view implausible and decided that his ether state of rest was physically superfluous structure, so that the two theories were really just variants of the same theory.

Natural and Cultured Pairs Similar stories play out for the other examples. The possibility of discounting the physical reality of superfluous structure leaves the physical distinctness of the pairs in question. Among the natural pairs, consider the many observationally equivalent formulations of Newton's mechanics, each with a distinct inertial state of motion designated as the true, absolute state of rest. A Newtonian can insist that the formulations differ on a matter of fact: the true disposition of absolute rest. However the overwhelming modern consensus is that this is the wrong way to understand Newtonian mechanics. The absolute state of rest plays no role in the deduction of observational consequences and, because of its perfectly symmetrical entry into each theory, we have no basis in evidence to decide which inertial motion it coincides with. So it is routinely discarded as physically superfluous and all the formulations regarded as physically equivalent.

A similar analysis arises in the case of the standard formulation of Newtonian gravitation theory in terms of gravitational fields and the Cartan curved space-time formulation. One could insist that the two are distinct in that the former posits a background set of inertial motions not present in the latter. Following the model of general relativity, it is now standard to assume that this background inertial structure is physically superfluous. Indeed, in the special case of homogenous cosmologies, symmetry arguments essentially similar to those used in the case of Newton's absolute space and Lorentz's ether makes insistence on the physical reality of the background inertial structure unsustainable (see Norton 1995).

Consider quantum mechanics and Bohm's theory, setting aside again that they are not strictly observationally equivalent. Bohm's theory adds a definite, hidden position for the particle, always possessed by it at every moment, and our ignorance of its true value is expressed in a probability distribution. A Bohm theorist can insist that this is a physically real addition to the ontology, so that the Bohm theory is physically distinct from traditional quantum mechanics. A traditionalist can reply, however, that the particle position only becomes manifest at the moment of measurement, so that standard quantum mechanics can assert that the position and its probability distribution came to be at the moment of measurement. All a Bohm theorist has done is to project the position and associated probability distribution back in time to the initial set up—a superfluous addition since all the theoretical information needed to specify the actual measurement outcome is already fully encoded in the wave function. The debate over the proper attitude to take to the Bohm theory is ongoing, and I stress that I do not want to take a side here. All I want to point out is that there are readily available arguments for the physical superfluity of the additional structure posited by the Bohm theory, so the example is not an unequivocal instance for use in the induction to the underdetermination thesis.

Among the cultured pairs, the best known are the Poincaré-Reichenbach cases of multiple geometries. One could insist that Reichenbach's universal force field is a physically real field so that the different geometries are physically distinct. No one, neither Reichenbach nor his critics, wanted to take the differences seriously, physically, and all regard the multiple accounts as simply variant presentations of the same facts. Reichenbach thought the examples demonstrated the conventionality of geometry and his critics thought they demonstrated the artificiality of universal force fields (see Norton 1992, §5.2.4). I am less sure of what the correct analysis is for the remaining cultured examples, observationally equivalent space-times and continua with and without reals. However, exactly because the examples are highly contrived, I am not inclined to read much significance into them for the import of evidence in real science.[13] A finitist in geometry might well want to read the addition of reals to rational continua as physically superfluous structure, so the two representations of continua would be merely variant formulations of the same physical facts. Alternatively, since the rationals fully fix the real insofar as the reals can be defined as Dedekind cuts of rationals, the example might illustrate what I shall call below "gratuitous impoverishment."

Additional Structure Comes to Be Judged Physically Superfluous

While we can always insist that pairs of logically distinct theories represent distinct facts, the historical tendency has been to reinterpret additional structure not essential to deriving a theory's observational consequences as physically superfluous. Such was the case with Newton's absolute space, Lorentz's ether state of rest, and the background inertial structure of Newtonian gravitation theory. It is not the case with Bohm's theory, which seeks to reinstate a type of ontology discarded by traditional quantum theory.

Sometimes the recognition that two competing theories are really variants of the same physical facts can be a cathartic release from a protracted debate in which neither side seems able to display empirical evidence that decisively favors their view. In special relativity, for example, there was a long debate over how heat and temperature transform under the Lorentz transformation, with several incompatible candidates advanced. It was eventually resolved with the decision that the differences corresponded to nothing physical (see Liu 1994). There was a similar debate over the correct Lorentz transformation for the energy and momentum of stressed bodies in special relativity. It was resolved with the decision that the different transformations depended on differences in a definition hidden in the formalisms (see Janssen 1995).

Gratuitous Impoverishment and the Artificial Pairs

The natural and cultured pairs of observationally equivalent theories can be construed as variant formulations of the same physical theory if we regard the additional structures of one or other of the pair as physically superfluous. The artificial pairs are similar insofar as one member of the pair does accord physical significance to a structure whereas the other member of the pair does not. In these artificial cases, however, I want to urge that the second deprives this additional structure of physical significance improperly. That is, it represents what I shall call a "gratuitous impoverishment" of the second theory. Unlike the natural and cultured pairs, in artificial pairs the additional structure is essential to both theories' derivation of their observational consequences and is well confirmed by these observational consequences. Indeed, in most cases the additional structure is fixed by the observational evidence, whereas the superfluous structure of the natural and cultured pairs is usually not. If we discard this additional structure, we lose an essential part of the machinery of both theories and deny something for which we have good evidence—the academic equivalent of burying one's head in the sand.

For example, a revisionist geology must specify that the world was created in 4004 BC complete with the fossils that would have been formed if the earth had the ancient history presumed by traditional geology. So the theoretical structures of traditional geology are essential to the formation of the revisionist rival, but the ancient past of that geology is then denied. Yet our fossil record is surely strong evidence for that ancient past; it is certainly no evidence at all for a creation in 4004 BC. Similarly, our experiences of the ordinary world are surely strong evidence for the reality of the ordinary world; they are certainly no evidence at all for the supposition that our brains sit in vats of nutrients and are fed simulated experiences.

The same remarks apply to Kukla's algorithm for generating an observationally equivalent rival T' for any theory T: construct the theory T' with identical observational consequences as T, but with the negation of all of the theoretical claims of theory T. Kukla offers the procedure as an algorithm for generating empirically equivalent rivals for any nominated theory, so that it amounts to a proof of the underdetermination thesis. If we assume that the algorithm is applied to a well-formulated theory T whose theoretical structure is essential to T's generation of observational consequences, then the construction of T' amounts to a gratuitous impoverishment of theory T, the denial of structures that are essential to the derivation of observational consequences and that are well confirmed by them.

Even if we set aside the concern of gratuitous impoverishment, the proof fails since it does not deliver an equally confirmed rival. That the world appears just as if theory T is true surely confirms more strongly the theoretical claims of T than it does their negations. Success of the proof requires us to suppose otherwise—that both the theoretical claims of T and their negations are equally confirmed in *every* case. Of the accounts of confirmation canvassed here only one, defective account can give us that.

The underdetermination thesis would prove to be something much less than we imagined if it merely asserts that the determining power of evidence is foiled by the utterly fantastic deception of these artificial pairs.

In conclusion, the underdetermination thesis offers a simple and profound restriction on the reach of evidence. Its appeal in science studies is strong, for it affords a simple argument that is both general and principled for the necessity of social and other factors in the determination of the content of scientific theories. Anything that easy and powerful seems too good to be true. And that is my principal claim: it is too good to be true; and the arguments in its favor have employed hasty stratagems that do not bear scrutiny. Our examinations of the nature of inductive inference have proven it to be sufficiently complicated to support no simple thesis that evidence must underdetermine theory—or, for that matter, that evidence must determine it. I see no easy escape. General claims on the relative weight of evidential and other factors in the determination of scientific theories will need to be supported by careful scrutiny of the particular cases at hand. There are no shortcuts.

NOTES

I thank participants in the First Notre Dame-Bielefeld Interdisciplinary Conference on Science and Values (Zentrum für interdisziplinäre Forschung, Universität Bielefeld, 9–12 July 2003) for helpful discussion. I am also grateful to Peter Machamer, Sandra Mitchell, and members of their NEH Summer Institute in Science and Values (23 June–25 July 2003, Dept. of HPS, University of Pittsburgh) for helpful discussion on an oral presentation of a preliminary versions. I also thank John Earman.

1. Everyone has their own favorite examples of the latter. Mine is described in Norton (2000, 72).

2. For an entry into this literature, see Longino (2002, 124–28; 40–41), from whom I take the word "gap," and Laudan (1990, 321). Use of the gap argument need not always invoke the underdetermination thesis by name. Bloor (1982) in effect introduces it with the

claim, drawing on Hesse's work, that the world is unable to determine a classificatory system for use in science, but that the stability of our theoretical knowledge comes "entirely from the collective decisions of its creators and users" (280).

3. In addition to the works cited below, see Ben-Menachem (2001); Boyd (1973); Brown (1995); Earman (1993); Magnus (2003); and McMullin (1995).

4. Other strands of criticism include Grünbaum's (1959) challenge that there is no assurance that nontrivial, alternative auxiliary hypotheses will be available to protect any given hypothesis from refutation, and that sometimes they are assuredly not available. Laudan and Leplin (1991) and Leplin (1997) object that the empirical equivalence that underpins the thesis depends essentially on the false presumption of fixity of auxiliary hypotheses.

5. One might wonder if Quine, the wellspring of the modern underdetermination thesis, is really advocating this trivial version when he writes: "Naturally [physical theory] is underdetermined by past evidence; a future observation can conflict with it. Naturally it is underdetermined by past and future evidence combined, since some observable event that conflicts with it can happen to go unobserved" (1970a, 178–79).

6. Is Quine committing this fallacy when he continues the preceding quote: "Moreover many people will agree, far beyond all this, that physical theory is underdetermined even by all *possible* observation" (1970a, 179)? Or is it another snake intentionally hidden in his stylistic garden?

7. While this view is often properly labeled a version of "holism," Quine (1975, 313) insisted that the term be reserved for Duhem's view that hypotheses could not be tested against experience in isolation.

8. For details of the adjustments needed, see Janssen (1995).

9. I say "lore" here since the situation with the original theories and proofs was not so simple. See Muller (1997).

10. As Bohm pointed out in his original paper, since his theory does not collapse the particle's wave function unlike ordinary quantum mechanics, there remains an extremely small probability of some observable effect arising from the uncollapsed portion; it is "so overwhelmingly small that it may be compared to the probability that a tea kettle placed on a fire will freeze instead of boil" (Bohm 1952, n18). Curiously the exact same remark could be used to dismiss the practicality of observations that could distinguish a phenomenological thermodynamics from a kinetic theory of heat. Yet such observations proved feasible once we stopped looking at intractable cases likes kettles on fires. Other possible observable differences have been discussed. See Cushing (1994, 53–55).

11. The simple argument is (briefly) that compensating adjustments to different hypotheses in a theory yield an alternative theory with the same observational consequences and is equally confirmed by them.

12. I have argued that this requirement of simplicity is not the injection of a mysterious metaphysics of simplicity into induction relations. Rather it is merely a somewhat blunt way of adding in further factual conditions known to prevail in the case at hand and which are decisive in determining the import of the evidence (see Norton 2003, §3).

13. The space-time example is at the contrived extreme of the cultured examples since it is the observations made by just one observer that fail to determine the space-time; the totality of observations of all observers does not, although we have no way physically to

collect all the observations at one event. Also the cosmology presumed is somewhat impoverished in that it is assumed that an observer can only have evidence for the space-time structure at an event if the observer can receive a light signal directly from it. Standard cosmological theories are rich enough in additional theory to fix routinely the space-time structure at events in the cosmos outside our past light cone.

REFERENCES

Ben-Menachem, Yemima. 2001. "Convention: Poincaré and Some of His Critics." *British Journal for the Philosophy of Science* 52:471–513.

Bloor, David. 1982. "Durkheim and Mauss Revisited: Classification and the Sociology of Knowledge." *Studies in History and Philosophy of Science* 13:267–97.

Bohm, David. 1952. "A Suggested Interpretation of the Quantum Theory in Terms of 'Hidden' Variables, I and II." *Physical Review* 85:166–93.

Boyd, Richard N. 1973. "Realism, Underdetermination, and a Causal Theory of Evidence." *Noûs* 7:1–12.

Brown, James R. 1995. "Intervention and Discussion: Underdetermination and the Social Side of Science." *Dialogue* 34:147–61.

Cushing, James T. 1994. *Quantum Mechanics: Historical Contingency and the Copenhagen Hegemony.* Chicago: University of Chicago Press.

Duhem, Pierre. 1954. *The Aim and Structure of Physical Theory.* Trans. Philip P. Wiener. Princeton: Princeton University Press.

Earman, John. 1993. "Underdetermination, Realism and Reason." In *Philosophy of Science, Midwest Studies in Philosophy,* vol. 18, ed. Peter A. French et al., 19–38. Notre Dame: University of Notre Dame Press.

Glymour, Clark. 1977. "Indistinguishable Space-Times and the Fundamental Group." In *Foundations of Spacetime Theories: Minnesota Studies in the Philosophy of Science,* vol. 8, ed. John Earman, Clark Glymour, and John Stachel, 50–60. Minneapolis: University of Minnesota Press.

———. 1980. *Theory and Evidence.* Princeton: Princeton University Press.

Goodman, Nelson. 1983. *Fact, Fiction and Forecast.* 4th ed. Cambridge, MA: Harvard University Press.

Grünbaum, Adolf. 1959. "The Duhemian Argument." *Philosophy of Science* 27:75–87.

Hoefer, Carl, and Alex Rosenberg. 1994. "Empirical Equivalence, Underdetermination, and Systems of the World." *Philosophy of Science* 61:592–607.

Howson, Colin, and Peter Urbach. 1989. *Scientific Reasoning: The Bayesian Approach.* La Salle, IL: Open Court.

Kukla, André. 1998. *Studies in Scientific Realism.* New York: Oxford University Press.

Janssen, Michel. 1995. *A Comparison between Lorentz's Ether Theory and Special Relativity in the Light of the Experiments of Trouton and Noble.* PhD diss., Department of History and Philosophy of Science, University of Pittsburgh.

———. 2002. "COI Stories: Explanation and Evidence in the History of Science." *Perspectives on Science* 10:457–522.

Laudan, Larry. 1990. "Demystifying Underdetermination." In *Scientific Theories, Minnesota Studies in the Philosophy of Science,* vol. 14, ed. C. Wade Savage, 267–97. Minneapolis: University of Minnesota Press. Repr. in *Philosophy of Science: The Central Issues*, ed. Martin Curd and J. A. Cover, 320–53. New York: Norton, 1998.

Laudan, Larry, and Jarrett Leplin. 1991. "Empirical Equivalence and Underdetermination." *Journal of Philosophy* 88:449–72.

Leplin, Jarrett. 1997. "The Underdetermination of Total Theories." *Erkenntnis* 47:203–15.

Lipton, Peter. 1991. *Inference to the Best Explanation.* London: Routledge.

Liu, Chuang. 1994. "Is There a Relativistic Thermodynamics? A Case Study of the Meaning of Special Relativity." *Studies in History and Philosophy of Science* 25:983–1004.

Longino, Helen. 1990. *Science as Social Knowledge: Values and Objectivity in Scientific Inquiry.* Princeton: Princeton University Press.

———. 2002. *The Fate of Knowledge.* Princeton: Princeton University Press.

Magnus, P. D. 2003. "Underdetermination and the Problem of Identical Rivals." *Philosophy of Science* 70:1256–64.

Malament, David. 1977. "Observationally Indistinguishable Spacetimes." In *Foundations of Spacetime Theories: Minnesota Studies in the Philosophy of Science*, vol. 8, ed. John Earman, Clark Glymour, and John Stachel, 61–80. Minneapolis: University of Minnesota Press.

Mayo, Deborah. 1996. *Error and the Growth of Experimental Knowledge.* Chicago: University of Chicago Press.

McMullin, Ernan. 1995. "Underdetermination," *Journal of Medicine and Philosophy* 20: 233–52.

Muller, F. A. 1997. "The Equivalence Myth of Quantum Mechanics." *Studies in History and Philosophy of Modern Physics* 28: Part I: 35–61, Part II: 219–47.

Newton-Smith, William H. 2001. "Underdetermination of Theory by Data." In *A Companion to the Philosophy of Science*, ed. W. H. Newton-Smith, 532–36. Oxford: Blackwell.

Norton, John D. 1992. "Philosophy of Space and Time." In *Introduction to the Philosophy of Science*, Merrilee H. Salmon et al., chap. 5. Englewood Cliffs, NJ: Prentice Hall. Repr. Indianapolis: Hackett, 1999.

———. 1993. "The Determination of Theory by Evidence: The Case for Quantum Discontinuity 1900–1915." *Synthese* 97:1–31.

———. 1995. "The Force of Newtonian Cosmology: Acceleration Is Relative." *Philosophy of Science* 62:511–22.

———. 2000. "How We Know about Electrons." In *After Popper, Kuhn, and Feyerabend: Recent Issues in Theories of Scientific Method*, ed. Robert Nola and Howard Sankey, 67–97. Dordrecht: Kluwer.

———. 2003. "A Material Theory of Induction." *Philosophy of Science* 70:647–70.

———. 2005. "A Little Survey of Induction." In *Scientific Evidence: Philosophical Theories and Applications*, ed. Peter Achinstein, 9–34. Baltimore: Johns Hopkins University Press.

Quine, Willard V. O. 1951. "Two Dogmas of Empiricism." *Philosophical Review* 60:20–43. Repr. in *From a Logical Point of View*, Cambridge, MA: Harvard University Press, 1953.

————. 1970a. "On the Reasons for Indeterminacy of Translation." *Journal of Philosophy* 67:178–83.

————. 1970b. "Natural Kinds." In *Essays in Honor of Carl Hempel*, ed. Nicholas Rescher, 1–23. Dordrecht: Reidel. Repr. in *Grue!: The New Riddle of Induction*, ed. Douglas Stalker, 41–56. La Salle, IL: Open Court, 1994.

————. 1975. "On Empirically Equivalent Systems of the World." *Erkenntnis* 9:313–28.

Salmon, Wesley C. 1975. "Confirmation and Relevance." In *Induction, Probability and Confirmation: Minnesota Studies in the Philosophy of Science*, vol. 6, ed. Grover Maxwell and Robert M. Anderson, 3–36. Minneapolis: University of Minnesota Press.

————. 1984. *Scientific Explanation and the Causal Structure of the World*. Princeton: Princeton University Press.

2

VALUES AND THEIR INTERSECTION
Reduction as Methodology and Ideology

MARGARET MORRISON
University of Toronto

Typically when we think about values in science we think about the kinds of sociopolitical interests—what Longino (1986) calls contextual values—that sometimes negatively influence both the practice of science and its results. The consensus is that neither the political agendas of governments nor the discriminatory biases of societies should in any way determine the kind of research deemed worth pursuing; yet they sometimes do. It is generally agreed that allowing such interference results in bad scientific methodology, and compromises what we consider the objectivity of scientific knowledge and practice.

Other kinds of values, sometimes referred to as cognitive values, are seen as an essential part of science and embedded in its practice. These include the pursuit of explanatory power, simplicity, accuracy, and so on; values that help to define not only the goals of science but the way in which those goals are achieved. These Longino refers to as "constitutive" values. Both she and Kuhn (1977) have helped to clarify the distinction between constitutive and the more politicized contextual values, and the role each plays in both promoting and hindering scientific inquiry. Kuhn claims that while cognitive values function as objective criteria for theory choice, the way they are interpreted—that is, what is encompassed by each and how they are weighted against each other—will be determined by subjective factors specific to scientific groups and paradigms. But the important point here is that it is scientists themselves who make these decisions, not those who are external to the practice.

What I want to do here is not so much to challenge the Kuhn/Longino account of scientific values but to present a somewhat different picture of the way that sociopolitical values and scientific values can interact. The focus of my discussion is on reduction and the role it plays in evaluations among scientists themselves about what counts as acceptable practice.

Reduction is interesting for a number of reasons. First, it underlies many of our accounts of explanatory power and simplicity and is often thought to be the goal of scientific inquiry on a more general level. In other words, what we want from our theories is an understanding of the world we live in, an understanding that appeals to a few basic laws or principles. While this seems desirable, it is also a goal that is frequently questioned not only within the physics community (for example, can all of physics be ultimately understood in terms of high energy physics and a grand unified theory?) but also by biologists and social scientists, who argue that the world is simply too complex to be understood in reductionist terms. For example, one cannot simply reduce biology to physics by assuming that the laws and entities of one domain can serve to explain phenomena in the other.[1]

A related concern of biologists and social scientists is the applicability of reduction as the appropriate methodology for dealing with the kinds of entities they are concerned with: living organisms and populations. But this is not just an epistemological issue about which problems are best suited to reductionist analysis, or whether reduction can provide the kinds of answers we seek. There is a broader methodological concern that indicates a kind of moral or political unease. The core of the worry is that reduction is thought to embody a worldview that disregards the specific features of phenomena or individuals that may be important for understanding how they function and interact. Put differently, reduction assumes a kind of atomism that ignores complexity inherent in different individuals/phenomena and results in a sterile and inaccurate account of biological individuals and the societies in which they live. Allowed to gain credibility, such an approach can result in attitudes that ignore individual natures and treat living things in a generic, mechanical fashion. This in turn may result in undesirable social and political practices. In essence then, what began as a cognitive value has taken on a social dimension because of the way in which the methodology of reduction is revealed to have moral and political implications. Hence, reduction becomes a point of intersection in the debate regarding the role of cognitive and contextual values.

The example I want to focus on here is the way in which decisions about the use of reductionist methods in the development of population genetics changed the focus of certain aspects of evolutionary theory away from biological individuals, and in doing so created a new theoretical framework for explaining evolutionary processes. This particular story, more than perhaps any other I can think of, is infused with the overlap of cognitive and social or contextual goals, not only among the founders of population genetics and their predecessors but also among biologists who reacted and continue to react to the implementation of what they see as reductionist practices. The two who arguably played the largest role in the mathematization of biology are Karl Pearson (1857–1936) and Ronald A. Fisher (1890–1962), both staunch eugenicists who had a keen interest in formulating a theoretical foundation for the kind of biological theory capable of supporting their eugenic goals. Although sociopolitical concerns and scientific methodology were closely intertwined in their work, I want to claim that it is possible to clearly delineate a *scientific* source in the mathematization of biology and hence to separate methodological developments from the social concerns of its founders.[2]

Several biologists, including most prominently Ernst Mayr, have criticized the role played by mathematical population genetics for what they see as the neglect of the individual for the purposes of mathematical expediency. However, by focusing on the presuppositions behind both the methodology of mathematical/statistical biology and later criticisms of its reductionist methods, we can begin to see that many of these criticisms are themselves founded on antireductionist ideology rather than reasons that are properly scientific. So, while the objections are presented as scientifically based, a closer examination reveals a somewhat different motivation. But in order to illustrate the differences between what I call "scientific" objections and those that are ideologically based, we need to look at the ways in which biology *became* mathematical/statistical and the reasons behind the use of certain techniques. In particular, I want to illustrate how the motivation behind the methodologies used by Pearson and Fisher can be traced to attitudes regarding the use of statistics and mathematics for establishing the proper foundation of a scientific theory. While they were both enthusiastic eugenicists, these allegiances can be separated from their views about what constitutes the proper methodology for science.

Mathematizing Biology: Heredity as a Statistical Concept

The first to apply statistical methods to the study of heredity was Francis Galton (1822–1911). Although government agencies had been tabulating vast amounts of quantitative information about the population, there was no established statistical theory or methods capable of analyzing this data. Much of his early work (1865; 1869; 1885; 1889) was dedicated to showing that natural talents such as intelligence were inherited and followed a Gaussian distribution or the law of error. Although Galton followed Quetelet in applying the law of error to humans, his views diverged importantly when considering the interpretation or implication of error. Instead of focusing on the mean and ignoring variability as Quetelet did, error was exactly what Galton wanted to preserve since variability was seen as crucial for understanding the mechanisms involved in inheritance. He went on to develop the concepts of regression and correlation and formulated a law of ancestral heredity that described the inheritance of traits from one's ancestors as a decreasing geometric ratio.[3] Because Galton had no notion of multiple correlation and regression, his law gave incorrect results. It was Karl Pearson (1895) who was responsible for developing these latter techniques; he saw Galton's work not as "a biological hypothesis, but the mathematical expression of statistical variates . . . [which] can be applied . . . to many biological hypotheses" (Pearson 1930, 21).

The law of ancestral heredity became for Pearson a purely statistical formula for predicting the value of a trait from ancestral values. In other words, Pearson adopted a purely instrumental, phenomenological interpretation of the notions involved in heredity. Neither the correlation of a somatic character in a great-grandparent and grandchild nor the corresponding regression coefficient was a measure of what the former generation actually contributed to the latter. Instead, what was important was what could be predicted of somatic characters of the offspring from what the germ plasms had produced in the past. Hence, "contribution of an ancestor" was understood solely in terms of the value it contributed to the formula for predicting the offspring. Pearson claimed that the law of ancestral heredity was simply a statement of a fundamental theorem in statistical theory of multiple correlation applied to a particular type of statistics. If the statistics of heredity are themselves sound, the results deduced from the theorem will remain true regardless of which biological theory of heredity is in place.

Pearson's phenomenalism was motivated in part by his beliefs about the nature of individuals and how they should be represented in the context of a proper statistical analysis. In *The Grammar of Science* he claimed that no two physical entities are exactly alike; instead they form a class with variation about a mean character. Hence, even in physics the ultimate basis of knowledge is statistical, and the notion of sameness applied to molecules is only statistical sameness (Pearson 1900, 156). Because the category of causation seems to require absolute sameness, something that does not hold even in science, it should be replaced with the statistical idea of a correlation between two occurrences, which is, in effect, capable of "embracing all relationships between absolute independence to complete dependence."

Science itself could not survive describing only individual experiences; instead its conclusions are based on average experiences, no two of which exactly agree. The variability that is characteristic of experience may be attributed to errors in observation, impurities in specimens, physical factors in the environment, and so on. But when these are removed by a process of averaging, one passes from the perceptual to the conceptual and so from the real world to a "model" world (Pearson 1911, 153). So not only do we average over the individuals that constitute the population, but we also average over our experiences. The process of statistical analysis then involved the construction of models that could, in the case of biometry, be used to predict the inheritance of certain traits from parent to offspring. This "model" world had two important features: it was constructed by a process of averaging, and it involved populations much larger than those that could be investigated using experimental means. Large populations were necessary if biometrical statistics was to successfully establish its conclusions. And if the samples were large enough, then one could supposedly substitute the sample statistics for the population parameters.[4] These statistical populations would become the object of inquiry for the biometricians, with the accompanying notions of heredity and selection defined in purely statistical terms.

Pearson went on to write a series of eighteen papers between 1894 and 1912 entitled "Mathematical Contributions to the Theory of Evolution," which resulted in the mathematization of Darwinism and the beginnings of a move away from the kind of naturalist investigations that had characterized biology. The goal was to establish natural selection on a firm scientific foundation that would provide the basis for its application to human societies. This was the continuation of a process that began with Galton, both in terms of the development of statistics and the way statistical Darwinism could be put to use for eugenic purposes: "The general conclusion one must be forced to make, by accepting Galton's theories, is the imperative importance of humans doing for themselves what they do for cattle, if they wish to raise the mediocrity of their race" (Pearson 1899, 34).

Following along on the idea, Weldon and Pearson wrote in the editorial to the first edition of *Biometrica*: "The recent development of statistical theory, dealing with biological data on the lines suggested by Mr. Francis Galton, has rendered it possible to deal with statistical data of a very various kind in a simple and intelligible way, and the results already achieved permit the hope that simple formulae, capable of still wider application, may soon be found" (*Biometrica* 1 [1901]: 1-2). As Galton himself remarked in the same publication, "The primary object of Biometry is to afford material that shall be exact enough for the discovery of incipient changes in evolution which are too small to be otherwise apparent."

The crucial issue here is how Pearson's faith in and reliance on statistical methodology affected his views about both the methodology and content of Darwinism. For Pearson, blending inheritance and selection could only be established using statistical methods, since that was the only route to reliable knowledge. A restriction of variability and a modification of the mean character were important indices of a selective process. So, deviations from the populational mean type needed to be measured; that is, small innumerable variations were now defined in terms of a frequency distribution. If natural selection existed, it should affect the frequency distribution of the character on which it acted. In order to demonstrate natural selection, the mortality rate had to be shown to depend on variations in a given character; and so by accumulating comparable data on the correlation between mortality rate and variation, natural selection could be established on the basis of statistical inference.

Using the theory of multiple correlation, Pearson developed what he termed a "fundamental theorem" that involved correlated characters measured in successive generations of the same population. This made it possible, in principle, to predict the character of the offspring given an exhaustive knowledge of the character's distribution among ancestors. He went on to distinguish natural selection from what he termed reproductive selection, which was an analogue of human artificial selection, something that he thought would be one of the most powerful modification forces. The theory of selection became then a theory about a plurality of interacting agencies. Here Pearson's eugenics becomes important because natural selection was thought to no longer guarantee progress in civilized humanity; there was simply no correlation in the upper classes between viability and fertility. At the heart of this was the power of statistics as a predictive tool. As Pearson remarked to Galton, "the problems of evolution were in the first place statistical, in the second place statistical, and only in the third place biological" (Pearson 1930, vol. 3A, 128).

Pearson's goal was to redefine the problem of selection as a problem about heredity, which he could then treat from within the theory of correlation. In the hands of Pearson, heredity was nothing other than a correlation coefficient— its true measure was the numerical correlation between some characteristic or organ as it occurs respectively in parent and offspring.[5] Although Mendel's laws were rediscovered in 1900, Pearson opposed Mendelism as a system of heredity for two related reasons.[6] The first was, quite simply, its use of genes (Mendelian factors) as an explanation of heredity. Pearson saw the goal of science not as the postulation of causes but description of facts using statistical methods. The second concerned what he perceived as the inability to incorporate the Mendelian hypothesis into the statistical framework of biometry. The mathematics required for establishing conclusions involving anything beyond one or two Mendelian factors was intractable; too much information was required to take proper account of all the relevant variables. This intractability was eventually overcome by R. A. Fisher, but not before he had radically transformed the way mathematics and statistics were applied to biological questions.

We can see that Pearson's work represents in some sense a reduction of Darwinism to statistics, and with this a focus on large, mathematically constructed populations—statistical models, if you will—that replace the kind of natural populations studied by field biologists, and indeed those referred to by Darwin himself. But this move away from the notion of the individual as the object of inquiry was grounded in well-defined methodological principles. Biology, by nature, had to be statistical. Not only did Pearson think that science could not proceed by taking account of knowledge of individual cases as described in *Grammar*, but biological relationships could not be characterized in a mathematically exact and deterministic way: "The causes in any individual case of inheritance are far too complex to admit of exact treatment" (Pearson 1896, 115). He remarked in a letter to Yule that while there might possibly be a single-valued function relation in biology as in physics, this would require taking account of thousands of variables, and no approach to such a relation is even attempted when making observations. Hence one's investigations should begin with a frequency surface rather than a simple relation, and if a regression line was sought then one must find the line appropriate to the surface.[7]

That this work was coincident with his eugenic goals is not to suggest that the latter shaped the former. Pearson's statistical biology was firmly rooted in his views about the proper methodology for science and how to best establish natural selection as a scientific fact. It was only by accomplishing the latter goal that he could hope to put Darwinian theory to work in the task of improving the human race. So, while the cognitive and contextual values were mutually reinforcing in that the science could and should be put to social use, the statistical methodology clearly had a life of its own—one that was firmly rooted in how science *ought* to be done.

Increasing Abstraction: Infinite Populations of Genes

With the work of R. A. Fisher, biometrical statistics became embedded in a more mathematically complex framework appropriate for dealing with questions about the causes of heredity. Fisher's synthesis of Darwinism and Mendelism contributed to a further redefinition of the statistical aspects of selection, which in turn was accompanied by an even greater move away from the individual as the object of investigation and explanation. While Fisher had no obvious objections to Mendelism as a theory of heredity, what he was concerned about was the issue of prediction. Mendelism was capable of predicting with certainty the possible types of children of given parents; biometry, on the other hand, was more vague but capable of wider application. The probable measurement of particular characters of the offspring could be calculated from those of the parents and those of the general population, but large numbers of families of similar parents in that population were required before the prediction would be accurate. Fisher criticized the Mendelians for their preoccupation with detail and lack of attention to abstract reasoning; he claimed that no amount of information about the mechanism of heredity could do away with the need for broad coordinating principles of evolution of the kind provided by biometry (Fisher 1911, 60). What was needed was some way of incorporating a Mendelian model of inheritance and statistical methods similar to those used by the biometricians.

Fisher's genius was to use the methodology of statistical mechanics as a way of modeling populations whose hereditary traits resulted from combination of large numbers of Mendelian genes. The comparison was between populations redefined in terms of Mendelian factors (genes) and the populations of molecules that constitute a gas. The agencies of selection always act amid a multitude of random causes, each of which may have a predominant influence if we fix our attention on a particular individual. Yet, these agencies determine the progress or decline of the population as a whole. In the case of the kinetic theory we have molecules moving freely in all directions with varying velocities, yet we can obtain a statistical result that is a perfectly definite measurable pressure. Knowledge of the nature and properties of the atom is inessential and independent of our knowledge of general principles in the way that our ability to predict and control the way populations evolve is independent of particular knowledge of individuals (Fisher 1911). This analogical model would loom large in Fisher's later work, informing the way in which biological populations should be conceived of in order to produce the appropriate kinds of statistical results.

But why did Fisher think that the kinds of probabilities associated with kinetic theory would be appropriate for modeling populations of genes? The answer comes from the way in which he conceived of a Mendelian system and his views about what kinds of probabilities were appropriate for scientific inference.[8] Fisher's gas theory analogy suggested a new way of conceptualizing populations of genes, which was significantly different from the way biometric analysis attempted to characterize populations. Pearson's requirement regarding the amount and kind of information needed to describe a population made it impossible to incorporate anything more than two or three Mendelian factors; the mathematics was simply too cumbersome. By introducing sophisticated diffusion techniques and relaxing the kinds of assumptions required for specifying the population structure, Fisher was able to achieve a synthesis of Darwinism and Mendelism that was simply impossible for Pearson, even if it had been his goal. This synthesis produced a statistical interpretation of selection that was different from Pearson's, not only because it included the notion of a gene, but because the population structure was different from the kinds of populations considered by the biometricians. In order to fully understand the story, we need to look at Fisher's views about the kind of statistical model that was exemplified in Mendelism and how its union with the kinetic theory was significant for this reconceptualization of Darwinian selection.

In Darwinian theory, chance was involved in the origin of inherited variation, but this was because we had imperfect knowledge of the causes that governed these events, not because the causes of variation were random. Because the biometricians eschewed any appeal to causes, the origin of variation was not taken as an essential feature that one needed to incorporate into an account of heredity. Instead the phenomenon of heritable variation was expressed in terms of statistical correlation. With Mendelism, however, the situation was more like tossing a coin; it involved a chance mechanism that generated with probability 1 a set of clearly defined outcomes (Howie 2002). To that extent, the genetic probabilities could be considered objective features of the world rather than a measure of ignorance. Because the gametes distribute by chance, the long-run frequency of a specific genotype in a large offspring generation can be predicted exactly from the genetic makeup of the parents and the rules of combinatorial analysis. Probabilities were defined as frequency ratios in a hypothetical infinite population from which the actual data are a random sample; so probabilities were used to calculate sample distributions from populations.

A probability, like the frequency of a gene, was definable in terms of a ratio in a population. In genetics, prior probabilities (assessments of the distribution of these populations) could be generated objectively by combinatorial analysis of an individual's lineage. Because Mendelian probabilities dealt with the distribution of gametes, they did not change with new information. Fisher drew a distinction between likelihood, defined as the confidence that a particular parameter occurred in the population given the sample, and probability defined in terms of relative frequency.[9] Likelihood referred to an inference made about the value of a particular probability distribution for a specific sample. In other words, likelihood applied to hypotheses inferred about the data, and probability referred to observed frequencies. As Fisher himself remarked, we know nothing of the probability of hypotheses, and to speak of likelihood as an observable quantity is meaningless. Just as Pearson's statistical thinking defined what was appropriate scientific practice, so too with Fisher. Mendelism generated a model of probability as well as a methodology for evaluating hypotheses. According to Fisher, one of the difficulties with biometrical methods was the conflation of the statistics of empirical surveys with the parameters of the underlying population. The problem of estimating one from the other was a matter of inference; the larger the sample, the better the approximation to the complete population.

As Pearson was aware, putting selection on a firm footing required a sophisticated statistical analysis that allowed one to average over individual variations to draw general conclusions about a population. But, once one takes account of genes as the mechanism of heredity then the constitution of the "population" changes from individuals, in the Darwinian and Pearsonian senses (statistically defined), to genes. Consequently, a different kind of statistical methodology is required to characterize the ways in which selection would operate in populations of genes. As in the case of the velocity distribution law, where a sufficiently large number of independent molecules would exhibit a stable distribution of velocities, a sufficiently large Mendelian population should enable one to establish general conclusions about the presence of particular traits.

Fisher's goal was to determine the extent to which characteristics such as stature were determined by a large number of Mendelian factors (Fisher 1918). Studies had shown that in the case of brothers the correlation coefficient is around .54 (amount of variance due to ancestry), which leaves 46 percent of the variance to be accounted for in some other way. According to the Mendelian hypothesis, the large variance among children of the same parents is due to the segregation of those factors in respect to which the parents are heterozygotes. Fisher wanted to separate how much of the total variance was due to dominance, how much resulted from other environmental causes, and how much from additive genetic effects. If one could resolve observed variance into these different fractions (by expressing these fractions as functions of observed correlations) then one could easily determine the extent to which nature dominated over nurture. He succeeded in showing that the effect of dominance in individual effects expressed itself in a single dominance ratio. Using fraternal correlation, Fisher was able to determine the dominance ratio and distinguish dominance from all nongenetic causes such as environment (which might possibly lower correlations). Essentially what Fisher did was distinguish not only between genetic and environmental variance but also between the different components of genetic variance itself (additive and dominance).

But what exactly was it that enabled Fisher to perform this kind of statistical analysis? He made a number of explicit assumptions that were clearly at odds with Pearson's approach (1904, 1909b). Perhaps the most important difference was the assumption of an indefinitely large number of Mendelian factors that had been perceived mathematically intractable and out of the region of experiment using Mendelian methods. But such an assumption was necessary to guarantee Fisher's statistical results. However, for Pearson there were other factors that were crucial for a proper characterization of Mendelian populations. In particular, for each Mendelian factor one needed to know which allelomorph was dominant; to what extent dominance occurred; what were the relative magnitudes of the effects produced by different factors;

in what proportion did the allelomorphs occur in the general population; were the factors dimorphic or polymorphic, to what extent were they coupled, and so on.

In addition there were the more general considerations that needed to be taken into account regarding homogamy (preferential mating) as opposed to random mating, selection, and environmental effects, all of which needed to be treated separately if one was to determine the genetic basis of the inheritance of particular characteristics. If one assumed, as Fisher did, an indefinite number of Mendelian factors then the nature of the population could not be specified in any complete sense, thereby undermining any statistical result that might follow.

Contra Pearson, Fisher did not assume that different Mendelian factors were of equal importance, nor did he assume that all dominant genes had a similar effect. In order to simplify his calculations, Fisher also assumed random mating as well as the independence of the different factors. Finally, the factors were sufficiently numerous that some small quantities could be neglected. So, not only did Fisher differ from Pearson with respect to specific assumptions about the *nature* of Mendelian factors (that all were equally important, etc.), but assumptions necessary to characterize a Mendelian population were much more general. By assuming an indefinite number of factors, it was possible to ignore individual peculiarities and obtain a statistical aggregate that had relatively few constants.

Once the causes of variance were determined, Fisher went on to specify the conditions under which variance could be maintained and how gene frequencies would change under selection pressures and environmental conditions (Fisher 1922). To answer these questions he introduced stochastic considerations and examined the survival of individual genes by means of a branching process analyzed by functional iteration, and then set up the "chain-binomial" model and analyzed it by a diffusion approximation. Fisher also examined the distribution of factors not acted on by selection and cases of gene extinction counterbalanced by mutation. He was able to show that even in a population of roughly ten thousand random-breeding individuals without new mutations, the rate of gene extinction was extremely small. Hence, the chance elimination of genes could not be considered more important than elimination by selection.

The other important component in the analogy with gas theory was the fact that the distribution of the frequency ratio for different Mendelian factors was calculable from the fact that the distribution was stable in the absence of selection, random survival effects, and so forth. Again, the source was the velocity distribution law in gas theory. Just as the formulation of this law assumed an independence of molecules in the gas, so too Fisher assumed the independence of various hereditary factors from each other and their independence from the effects of selection and random survival. He showed that the presence of even the slightest amount of selection in large populations had considerably more influence in keeping variability in check than did random survival. Consequently, the assumption of genotypic selection balanced by occasional mutations fit the facts deduced from the correlations of relatives in humans.

So, by making simplifying assumptions about the large size of the population and its high degree of genetic variability, Fisher was able to demonstrate how his stochastic distributions led to the conclusion that natural selection acting on single genes (rather than mutation, random extinction, epistasis, etc.) was the primary determinant in the evolutionary process. He also found that mutation rates significantly higher than any observed in nature could be balanced by very small selection rates. Like Pearson, whose biometric methodology served as a model for the practice of science, so too with Fisher and his emphasis on Mendelism and the kinetic theory. If nature is a gambling machine of the sort defined by Mendelian ratios, then it is possible to generate data of repeatable and independent trials. The models of probability and inference embedded in Mendelism and kinetic theory defined, for Fisher, the methodology of good science. The distribution of the gene ratio provided the ultimate expression of selective effects, because the gene remains the only trace of the existence of an individual in a population.[10] Given that selection acts on the heritable, what is important is the mean effect of each allele; it is the genetic structure of the population taken as a cloud of particles that evolves. Although we cannot know the state of each of the genes in the population, we can know the statistical result of their interaction in the same way that gas laws can be deduced from a collection of particles. Indeed, all observable processes of evolution can be described in the language of the statistical distribution of genes. The natural population is simply an aggregate of gene ratios.

We can see how the development of population genetics has taken us away from explanations involving individuals to increasing levels of abstraction that focus on statistical and mathematical model populations. We have also seen why both Pearson and Fisher thought this necessary. In each case they had specific views about how to achieve predictive success and accuracy within a scientific framework. Pearson eschewed causal explanations and relied on statistical correlation to establish claims about inherited traits. Although Fisher subscribed to a Mendelian theory of heredity, he also criticized Mendelism for its lack of attention to what he called "abstract reasoning" of the kind necessary to ensure broad predictive and explanatory power. Both were eugenicists, but like Pearson, Fisher's methodology was developed not in conjunction with his political views but parallel to them. During the early twenties and thirties he wrote several papers for the *Eugenics Review* outlining factors he saw as crucial for the development of societies. Indeed, the later chapters in *Genetical Theory* (Fisher 1930) dealt with human fertility rates and the rise and decline of human civilizations. But, the important point is that the theory of statistics and the methodology for dealing with specific problems was motivated by the kinds of scientific concerns appropriate for treating the problem at hand. Once in place, methods like the analysis of variance could be applied to eugenic problems, but the choice of methodology was determined first and foremost by questions about how to analyze and measure variation.

As I mentioned above, the methodology introduced by the population geneticists, specifically Fisher, has come under criticism for its emphasis on mathematical abstraction, reduction, and the lack of focus on individuals. I also said at the outset that my intention is to question whether these criticisms are the result of ideological reactions to reduction as a way of dealing with biological phenomena, or whether there are scientific reasons for thinking that this approach fails to answer the kinds of questions appropriate for biological contexts. Although the criticisms are presented in the spirit of the latter, a closer analysis reveals that this may not be the case.

Reduction and Ideology

It is perhaps interesting to point out that criticisms of Fisher's approach began as early as 1916, when he submitted his paper on the correlation between relatives for publication. Pearson, who was one of the referees, objected to (among other things) the assumption of an indefinitely large number of Mendelian characteristics, which he claimed was out of the region of experiment using Mendelian methods. Although Pearson was entirely comfortable dealing with large populations (a cornerstone of biometric methods), the idea that the manifestation of a character was the result of an indefinitely large number of Mendelian factors was anathema. While he acknowledged that one could increase the number of factors under consideration from two to perhaps even four, the idea that these could increase indefinitely was something he took to be amenable to neither proper statistical/biometrical analysis nor experimental tests of the kind favored by the Mendelians. The other referee, the biologist Punnett, also focused on this assumption, pointing out that although there were cases worked out for three factors, the mathematics proved very laborious, making it unlikely that Fisher's hypotheses could ever be tested by experiment. He puts the point rather strongly in a remark comparing this kind of work to problems that deal with "weightless elephants on frictionless surfaces, where at the same time we are largely ignorant of the other properties of the said elephants and surfaces" (Norton and Pearson 1976, 155).

Although the assumption regarding the large number of Mendelian factors was one that neither the Mendelians nor the biometricians saw as representing a realistic possibility, the objections were motivated by concerns about whether such abstract methods could yield the right sort of concrete results. In other words, they were methodological objections rooted in scientific concerns. Compare that with Mayr's remarks that treating population evolution as mere changes in gene frequency (as an input or output of genes) is like beanbag genetics, involving the addition and removal of beans from a bag (Mayr 1959). Here Mayr echoed an earlier statement by Waddington (1957), who claimed that the mathematical theory of evolution has not led to any noteworthy quantitative statements, nor did it reveal any new types of relations or processes that could explain phenomena that were previously obscure. In view of the earlier discussion, these comments seem not only unfounded but empirically false. What would motivate this kind of criticism in light of the empirical success of population genetics in unifying Mendelism and Darwinism? The answer, I believe, can be found by looking at some of Mayr's remarks regarding the distinction between typological versus population thinking. What these remarks reveal is a suspicion about the kind of essentialism that is associated with reductionism; an essentialism he sees as linked to undesirable practices concerning racial classification (among other things).

In the context of distinguishing essentialism, characterized by an interest in types, from population thinking, which stresses the importance of individuals, Mayr claims that "he who does not understand the uniqueness of individuals is unable to understand the working of natural selection" (Mayr 1982, 47). This is a strong claim indeed, especially since those who, by virtue of their methodology, seem to fall into the essentialist camp include the early population geneticists (especially Fisher), many of whom were avid Darwinians. Initially it seems peculiar to criticize these early geneticists for subscribing to methods more in keeping with typological as opposed to population thinking, since their focus was on an analysis of Mendelian populations. Mayr, together with Dobzhansky, characterize population thinking as promoting the genetic diversity of populations rather than homogeneity; and because of the tremendous statistical differences among populations we need to pay attention to individuals rather than simply associating them with particular populations defined homogeneously. Much of Mayr's criticism of the essentialist program focuses on the use and interpretation of mathematics in characterizing a population, where the notion of a population is understood as a "construct," dealing with the "average man" and variation around mean values. As an alternative, population thinking emphasizes the uniqueness of biological individuals; an approach that employs mathematics only as a way of representing the aggregate of a natural population where variation differs from character to character. In fact, Mayr goes so far as to say that "Darwin would not have arrived at a theory of natural selection if he had not adopted population thinking" (1982, 47).

Although Mayr never explicitly identifies the population geneticists and essentialism, one only need look at his characterization of essentialism to see quite clearly that the methods of the mathematical population geneticists and to some extent the early biometricians fall squarely within this camp. According to him, the essentialist treats biological individuals in the same way that physicists treats inanimate matter, constructing mathematical representations in order to calculate the characteristics of the average man and thereby determine his essential nature. Variation refers not to individuals but to "errors" around mean values. By contrast, the population thinker stresses the uniqueness of everything in the organic world and claims that we must approach these biological groups in a very different way than groups of inorganic entities. It is differences between individuals in natural populations that are real, not the constructed mean values calculated by the essentialists. Both Pearson and Fisher dealt with large statistical populations, largely neglecting actual populations of living organisms studied in the field or the laboratory. In addition, the emphasis on genetic selection and the attribution of a definite fitness value to each gene is seen by Mayr as un-Darwinian, leaving most of evolutionary phenomena unexplained (Mayr 1982, 588). In other words, the mathematical populations characterized by the geneticists actually prevent us from understanding the workings of natural selection.

Part of Mayr's criticism rests on the claim that in much of evolutionary biology the contributions of mathematics are very minor (1982, 41). Attempts to translate qualitative biological phenomena into mathematical terms have sometimes proven to be a complete failure because they lose touch with reality (1982, 54). For example, he accused the mathematical population geneticists (specifically Fisher) of oversimplifying, for the sake of mathematical tractability, the factors that entered their formulas. This led to a stress on absolute fitness values of genes, an overvaluation of additive gene effects, and to the assumption that genes rather than individuals are the target of natural selection. In summary, this kind of methodology has invariably led to unrealistic results (1982, 41). The more simplified the assumptions, the less one is able to account for the increasing complexities of biological phenomena; to that extent, simplification was intended to serve mathematical rather than biological ends. So mathematics, reduction, and abstraction are all considered tools of the essentialist, tools that are used inappropriately in the investigation of biological phenomena. It is worth noting that even contemporary authors share Mayr's views, to some extent. In a new afterword to his classic book, *The Origins of Population Genetics*, Provine (2006) states that "the models of the 1930s, still widely used today, [are] an impediment to the understanding of evolutionary biology."

Although my discussion of Pearson and Fisher addressed their attempts to establish a scientific foundation for Darwinism and its eventual synthesis with Mendelism, in telling that story I did not speak directly to the charges levied by Mayr and others about the problems with mathematical abstraction and its general inability to account for important problems in evolutionary biology. Instead my focus was on the coincident development of statistics and biology against the backdrop of eugenics. In order to argue that claims about the unimportance of mathematics are motivated by ideological rather than scientific concerns, I need to show first that they are largely unfounded. In order to do that it is necessary to show the ways in which certain aspects of biology, in particular population genetics, require reduction and mathematical abstraction to establish their conclusions, conclusions that are themselves necessary for understanding the way populations evolve. My story about the details of Fisher's and Pearson's accounts of the role of natural selection was an attempt to do just that.

On a very basic level, we can think of the problem in the following way. The objects of study in population genetics are primarily the frequencies and fitnesses of genotypes in natural populations. Some population geneticists define evolution as the change in the frequencies of genotypes over time, something that may be the result of their differences in fitness. Or to put it more accurately, we can model evolution by assigning fitnesses to genotypes and then following evolution as a change in allele frequencies. The problem is that while genotype or allele frequencies are easily measured, their change is not. Most naturally occurring genetic variants have a time scale of change that is on the order of tens of thousands to millions of years, making them impossible to observe. Fitness differences are likewise very small, less than 0.01 percent, also making them impossible to measure directly. What this means is that although the state of a population can be observed, the evolution of a population cannot be directly studied.

Like most systems that are not directly accessible, one investigates them by constructing mathematical models of the evolutionary process, then comparing their predictions with the behavior or states of the system, in this case populations that can be directly observed. But because one cannot know the genetic structure of a species (not only would we need a complete description of the genome but also the spatial location of every individual at one instant in time, which would then change in the next instant) the models need to incorporate certain idealizing assumptions that ignore the complexities of real populations. They focus on one or a few loci at a time in a population that mates randomly or has a simple migration pattern. These assumptions are necessary for characterizing the problem in a way that renders it tractable.

We can also make more specific claims about the importance of mathematical techniques. We saw in the discussion of Fisher's work how methods from statistical mechanics were crucial in achieving the synthesis of Darwinism and Mendelism. And indeed, certain aspects of this synthesis would not have been possible without intricate mathematical analysis. For example, the condition that the heterozygote must have a fitness exceeding each of the two homozygotes is not one for which a mathematical analysis is necessary, especially in the case where there is a stable equilibrium of allelic frequencies with both alleles represented at positive frequencies under selective pressures only. However, the corresponding claim in a k-allele system as well as results in the multilocus theory cannot be made without appeal to sophisticated mathematical techniques (Ewens 2004). Indeed, many of the central ideas in evolutionary biology originate in or are justified by purely mathematical analysis: for example, the fact that even weak selection can drive evolution, the importance of population structure in evolution, and the significance of drift and its interplay with selection. Drift concerns the small random changes in genotype frequencies caused by variation in offspring number between individuals and, in diploids, genetic segregation. It operates on a time scale that is proportional to the size of the population; so in a species of a million individuals it takes roughly a million generations for genetic drift to change allele frequencies in any significant way. Because there is no conceivable way of verifying that genetic drift changes allele frequencies in most natural populations, our understanding of it is purely mathematical/theoretical. Indeed most of the results in the stochastic theory of evolution require mathematical analysis.

Some problems, however, do not fit well into the framework of population genetics; phenotypic evolution is a case in point. This is because both morphological and behavioral phenotypes do not generally exhibit the kinds of simple, regular patterns that occur in allele transmission. Because the assumptions required to connect morphology to a Mendelian process are both extensive and arbitrary, the models cease to have the kind of foundational quality of those discussed above. However, even these problems can be addressed in a purely analytical way, provided the systems of interest are well defined. This kind of approach originated with Price (1970, 1972), who formulated a theorem that is an exact statement of the relationship between differential reproduction, inheritance, and evolution. As such it provided the beginnings of an evolutionary algebra and the kind of foundation for phenotypic evolution that Mendel's laws furnishes for population genetics. These kinds of results are important for my overall argument because they show the relevance and importance of mathematical analysis for cases that lie outside the theoretical framework of population genetics, and it shows that, contra Mayr, the importance of mathematical theory goes above and beyond simply providing rigor for qualitative statements.

Given these rather obvious advantages, questions arise regarding the motivation for Mayr's criticisms. Since they do not seem to be grounded in empirical findings we need to look elsewhere for clues about their source. One such source is, I believe, Mayr's views on the connection between typological thinking and racism. As I mentioned above, the methods of the population geneticists fit squarely within the categories that define typological as opposed to population thinking. The former deals with averages that are merely statistical abstractions, while the latter focuses on individuals and variation as the only real categories. We saw that in the cases of both Pearson and Fisher, statistical averages were, for different reasons and in different degrees, the object of investigation. Mayr (1959) sees typological thinking as the basis for all racist theories insofar as the typologist claims that every representative of a race conforms to a particular "type" which separates them from other races. Here the notion of type is defined in statistical terms. The populationist, on the other hand, defines race simply in terms of an average difference between two groups of individuals that is sufficiently great to be recognizable on sight. Because every individual will score in some traits above and in others below the average for the population, the notion of an ideal type does not exist. Consequently races are more properly characterized as overlapping rather than expressing certain "ideal" characteristics. Hence, by focusing on individuals rather than types, we get a very different notion of what constitutes a race and consequently a very different account of the role of this concept in social contexts.

As we saw above, the methodologies of both Pearson and Fisher involved, as a central notion, the concept of statistical averages. Pearson saw this as the only legitimate way to establish scientific conclusions about individuals. Fisher's concern was averages of genes, not individuals, which led to a reconceptualization of the notion of a population and how one should represent it mathematically in order to show the power of selection as the driving force of evolution. Behind this statistical methodology was a strong commitment to eugenics on the part of both Pearson and Fisher. But, as we also saw, their sociopolitical or contextual values were not the determining factor in the development of their methodologies. Nevertheless, one can see and appreciate the source of Mayr's worry. The use of reductionist strategies to support eugenic or racist goals is without question an undesirable outcome. Yet arguments against the latter should not involve the misrepresentation of the successes and goals of mathematical/reductionist techniques in biological contexts. Nor is the potential for misuse an argument for banishing the methodology altogether. Just as Fisher's and Pearson's statistical and mathematical work engaged purely scientific problems, so too should criticisms of reduction and mathematical abstraction. Criticisms of these techniques ought to focus on *scientific* problems and the role of the techniques in solving those problems and not be not motivated by ideological concerns that have little or no bearing on scientific success or failure.

Mathematical reduction is not always a desirable goal in physics, let alone biology. But, in some cases it provides us with the kind of knowledge that is simply not possible using other tools, techniques, and methods. Criticisms of reductionism that focus on the use to which it *might* be put in achieving detrimental ends can be as harmful as those ends themselves. Evaluating the contributions that reductionist techniques have made requires that we fairly evaluate both successes and failures, while keeping in mind the points of departure between the scientific and the social.

NOTES

I would like to thank the Social Sciences and Humanities Research Council of Canada for research support.

1. There are several contexts in which the merits and hazards of reduction have been played out. A widely publicized one is the debate between Steven Weinberg and Philip Anderson regarding the decision about whether to build the superconducting supercollider, a particle accelerator that would supposedly enable physicists to discover elementary particles and laws that govern the structure of the universe. Anderson, who opposed the building, argued that the physics embedded in this largely reductionist endeavor was simply not important for much of the physics relevant to our daily lives (see Weinberg 2001). Other examples of reductionist controversies include the debate over the status of sociobiology as an acceptable explanation of human behavior and the more recent emphasis on evolutionary game theory as a way of modeling human interactions.

2. Aspects of that story and the interplay of eugenics, Darwinism, and Mendelism are well documented in the literature, so I will not rehearse them here; see, for example, MacKenzie (1981); Norton (1975a, 1975b) who argue for a more ideologically based interpretation.

3. The law stated that "the share a man retains in the constitution of his remote descendants is inconceivably small. The father transmits, on average, one-half of his nature, the grandfather one-fourth, the great-grandfather one-eight; the share decreasing step-by-step in a geometrical ratio with great rapidity" (Galton 1865).

4. Pearson himself argued in his work on the χ^2 test that the sample and population constants differed on average by terms of the order $1/\sqrt{n}$, where n is the sample size, with the difference tending to 0 as n became large. Small samples simply did not provide the proper basis for statistical work (E. Pearson 1938, 379).

5. See Gayon (1998) for an excellent account of Pearson's views on the relation between heredity and correlation.

6. For an extended discussion of Pearson's opposition to Mendelism see Morrison (2002).

7. These remarks are taken from letters found in the Pearson Archives and quoted in Stigler (1986).

8. For a full account of this see Morrison (2002).

9. Howie (2002) has an excellent discussion of these issues as they relate to the broader context of interpretations of probability in the early twentieth century.

10. The Mendelian mechanism ensures that although a population may be said to have continued existence, the individuals that comprise it do not. The variability that passes from one generation to the next through reproduction is related to but not identical to phenotypic variability.

REFERENCES

Ewens, Warren J. 2004. *Mathematical Population Genetics, 1. Theoretical Introduction.* New York: Springer.

Fisher, Ronald A. 1911. "Heredity, Comparing the Methods of Biometry and Mendelism." Paper read to Cambridge University Eugenics Society. Reprinted in Norton and Pearson 1976, 155–62.

———. 1918. "The Correlation between Relatives on the Supposition of Mendelian Inheritance." *Transactions of the Royal Society of Edinburgh* 52:399–433.

———. 1922. "On the Dominance Ratio." *Proceedings of the Royal Society of Edinburgh* 42:321–41.

———. 1924. "The Biometrical Study of Heredity." *Eugenics Review* 16:189–210.

———. 1930. *The Genetical Theory of Natural Selection.* Oxford: Clarendon Press.

Galton, Francis. 1865. "Hereditary Talent and Character." *Macmillan's Magazine* 12:157–66, 318–27.

———. 1869. *Hereditary Genius.* London: Macmillan.

———. 1885. "Regression towards Mediocrity in Hereditary Stature." *Journal of the Anthropological Institute* 15:246–63.

———. 1889. *Natural Inheritance.* London: Macmillan.

Gayon, Jean. 1998. *Darwinism's Struggle for Survival.* Cambridge: Cambridge University Press.

Howie, David. 2002. *Interpreting Probability, Controversies and Development in the Early Twentieth Century.* Cambridge: Cambridge University Press.

Kuhn, Thomas S. 1977. "Values, Objectivity and Theory Choice." In *The Essential Tension.* Chicago: University of Chicago Press.

Longino, Helen E. 1986. *Science as Social Knowledge.* Princeton: Princeton University Press.

MacKenzie, Donald A. 1981. *Statistics in Britain, 1865–1930.* Edinburgh: University of Edinburgh Press.

Mayr, Ernst. 1959. "Where Are We?" *Cold Spring Harbor Symposia on Quantitative Biology* 24, 1–14. Repr. in *Evolution and the Diversity of Life*, 307–28. Cambridge, MA: Harvard University Press, 1976.

———. 1973. "The Recent Historiography of Genetics." *Journal of the History of Biology* 6:125–54.

———. 1982. *The Growth of Biological Thought.* Cambridge, MA: Harvard University Press.

Morrison, Margaret. 2002. "Modelling Populations: Pearson and Fisher on Mendelism and Biometry." *British Journal for the Philosophy of Science* 53:39–68.

Norton, Bernard J. 1975a. "Biology and Philosophy: The Methodological Foundations of Biometry." *Journal of the History of Biology* 8:85–93.

———. 1975b. "Metaphysics and Population Genetics: Karl Pearson and the Background to Fisher's Multi-factorial Theory of Inheritance." *Annals of Science* 32:537–53.

Norton, Bernard J., and Egon S. Pearson. 1976. "A Note on the Background to and Refereeing of, Ronald A. Fisher's 1918 Paper 'On the Correlation between Relatives on the Supposition of Mendelian Inheritance.'" *Notes and Records of the Royal Society* 31:151–62.

Pearson, Egon. 1938. *Karl Pearson: An Appreciation of Some Aspects of His Life and Work.* Cambridge: Cambridge University Press.

Pearson, Karl. 1895. "Contributions to the Mathematical Theory of Evolution, II: Skew Variation in Homogenous Material." *Philosophical Transactions of the Royal Society of London A* 186:343–414.

———. 1896. "Contributions to the Mathematical Theory of Evolution, III: Regression, Heredity, and Panmixia." *Philosophical Transactions of the Royal Society* 187:253–318.

———. 1898. "Mathematical Contributions to the Theory of Evolution. On the Law of Ancestral Heredity." *Proceedings of the Royal Society* 62:386–412.

———. 1899. "Theory of Genetic or Reproductive Selection." *Philosophical Transactions of the Royal Society A*, 192:259–78.

———. 1900. *The Grammar of Science.* 2nd ed. London: Black.

———. 1902. "On the Fundamental Conceptions of Biology." *Biometrica* 1:320–44.

———. 1904. "Mathematical Contributions to the Theory of Evolution. XII. On a Generalized Theory of Alternative Inheritance with Special Reference to Mendel's Laws." *Philosophical Transactions of the Royal Society A* 203:53–86.

———. 1909a. "The Theory of Ancestral Contributions in Heredity." *Proceedings of the Royal Society B* 81:219–24.

———. 1909b. "On the Ancestral Gametic Correlations of a Mendelian Population Mating at Random." *Proceedings of the Royal Society B* 81:214–29.

———. 1911. *The Grammar of Science.* 3rd ed. London: Black.

———. 1930. *The Life, Letters, and Labours of Francis Galton.* 3 vols. Cambridge: Cambridge University Press.

Price, George R. 1970. "Selection and Covariance." *Nature* 227:520–21.

———. 1972. "Extension of Covariance Selection Mathematics." *Annals of Human Genetics* 35:485–590.

Provine, William B. 2006. *The Origins of Theoretical Population Genetics.* Chicago: University of Chicago Press.

Stigler, Stephen M. 1986. *The History of Statistics: The Measurement of Uncertainty before 1900.* Cambridge, MA: Harvard University Press.

Waddington, Conrad H. 1957. *The Strategy of the Genes.* London: Allen and Unwin.

Weinberg, Steven. 2001. *Facing Up: Science and Its Cultural Adversaries*, Cambridge, MA: Harvard University Press.

◈ 3

VALUES, HEURISTICS, AND
THE POLITICS OF KNOWLEDGE

HELEN E. LONGINO
Stanford University

MANY PHILOSOPHERS HAVE ARGUED THAT scientific theory choice is guided by so-called "superempirical" values such as simplicity, comprehensiveness, or unification. While supporting acceptance of hypotheses that go beyond mere summations or generalizations of data, such values are held nevertheless to be epistemic, because they are held to be either truth-indicative or definitive of scientific understanding. Differently situated critics of the sciences dispute the claim of modern Western science to be value-free and truth-driven. In the United States, the most sustained critique has been articulated by feminist scientists, historians, and philosophers of science. In these reflections, I will draw on this feminist work. But I think the general points have broader applicability and can help us think about critical engagement with the sciences from other positions.

Most feminist writing in and about the content and methods of the sciences has exhibited two common themes: a kind of basic anti-reductionism and an emphasis on practical dimensions of inquiry; or the extensions of science into the social and material world. These themes resonate with a rubric already in use in philosophy of science: the rubric of the superempirical—cognitive, epistemic, scientific, or theoretical—values (which label one uses depends on the claim being made for them). Feminist themes can be read not just as critique, but also as versions of positively expressed values that contrast with those traditionally invoked. The term "virtues" might be more apt, indicating that we are talking here of properties or qualities of a theory, model, or hypothesis. I will suggest later that the term "heuristic" is even better.

For now I will stick with virtues or values. Theoretical virtues or values are those qualities or properties of a theory, model, or hypothesis that qualify it as at least praiseworthy, but also plausible, and even worthy of acceptance, or whose absence qualifies the theory for suspicion or rejection. I will first talk about a set of these virtues, and explain what they recommend and how they differ from orthodox or mainstream virtues. Then I wish to raise the question: what about these virtues would qualify them as virtues for a specifically feminist inquiry? Surprisingly, it is the answer to this question that provides material for more general reflections. I will arrive at those by integrating the answer about feminist inquiry with my own views about knowledge and inquiry. This will set the stage for some final reflections on the politics of knowledge.

The virtues I have found endorsed or advocated in feminist science studies include empirical adequacy, novelty, ontological heterogeneity, complexity or mutuality of interaction, applicability to human needs, and decentralization of power or universal empowerment. While empirical adequacy is held in common by feminist and nonfeminist researchers, the remaining five contrast intriguingly with more commonly touted values of consistency with theories in other domains, simplicity, explanatory power and generality, and fruitfulness or refutability. Many philosophers of science have invoked these more traditional or orthodox virtues, although their invocation is not without its mainstream critics.[1]

The Feminist Alternatives

Empirical adequacy means agreement of the observational claims of a theory with data. This trait is valued by feminists critical of misrepresentations of gender or gender relations in traditional theories and models. It is valued equally by feminist and mainstream scientists. Empirical adequacy, however, (even when supplemented by a requirement that there must actually be empirical or observational consequences to compare with data) is not a sufficient criterion for theory choice because of the philosophical problem known as the underdetermination of theory by data.

The underdetermination with which I am concerned is produced by a semantic gap between most hypotheses and the observational data adduced in evidence for them. For example, the data relevant to hypotheses about elementary particle interactions, collisions, and disintegrations are described in language different from the language used to represent the putative interactions—we don't directly observe pions and muons and neutrinos—we observe what we take to be their traces or effects in detectors (whether bubble tracks in compressed gas, numerical sequences on data tapes, transient bursts of electric current in shielded dense liquids). There can be no formal relations of derivability or a priori specifiable statements of evidential relevance between hypothesis statements and descriptions of data. The gap is filled in by a lot of theory—theory about particles, theory about detectors, possibly more—on which physicists rely in asserting the evidential relevance of what they can observe to particular hypotheses about particle interactions. The data alone could be compatible with several quite different hypotheses—the data do not carry their evidential relevance on their sleeves, so to speak. Whether they are evidence for a particular hypothesis or model is an empirical matter, judged against the background of assumptions about instruments, about the way the world is, and about the ways it is to be known.

The brute formal fact of underdetermination is addressed in different ways by different philosophers (and also described differently than I do here). One solution is to appeal to additional qualities of a theory, the theoretical or cognitive or superempirical virtues manifested by the theory. [2]

Feminists endorse the virtue of novelty of theoretical or explanatory principle as protection against unconscious perpetuation of the sexism and androcentrism of traditional theorizing, or of theorizing constrained by a desire for consistency with accepted explanatory models. The novelty envisioned is not the novelty of discovery of new entities (like the top quark) predicted by theory but rather of frameworks of understanding. For example, some feminist scholars (Haraway 1986; Sperling 1991) have criticized the articulation of female-centered models of evolution by feminist primatologists (such as Zihlmann 1978 and Hrdy 1981) as remaining too much within the framework of sociobiology, and thus perpetuating (what they take to be) other noxious values of that theoretical approach. Novelty, thus understood, is contrary to the value of consistency with theories in other domains as described by Kuhn (1977) or its variant, as propounded by Quine and Ullian (1978), conservatism (that is, preserving as much of one's prior belief set as possible).

Novelty and empirical adequacy are somewhat formal requirements. The next two concern substantive aspects of theories or models, and different aspects of the anti-reductionist theme. Any theory stipulates an ontology, that is, it specifies what is to count as a causally effective entity in its domain. A domain that is ontologically heterogeneous is one with different kinds of entities. An ontologically homogeneous domain contains only one kind. Such a domain is simpler than a heterogeneous domain in that only the properties and behavior of the one kind need be accounted for in models of the domain. Any one member can represent any other (at least in essential respects). Feminists who endorse heterogeneity as a virtue indicate a preference for theories and models that preserve heterogeneity in the domain under investigation, or that at least do not eliminate it on principle. An approach to inquiry that requires uniform specimens, that is, ontological homogeneity, may facilitate generalization, but it runs the risk of missing important differences, as when the male of a species comes to be taken as paradigmatic for the species (as in "Gorillas are solitary animals; a typical individual travels only with a female and her/their young"), or when, via the concept of male dominance, males are treated as the only causally effective agents in a population. Feminist scholars have instead insisted on observing, recording, and analytically preserving differences in populations under study (Altmann 1974). Their embrace of heterogeneity extends beyond human and animal behavior, however, and is also invoked in the context of genetic and biochemical processes. Feminist researchers have resisted unicausal accounts of development in favor of accounts in which quite different factors play causal roles. They therefore emphasize the multiplicity of kinds of factors at all developmental levels, from within the cell to the whole organism (Keller 1983, 1995). Heterogeneity, in its opposition to homogeneity, or uniformity, is thus opposed to ontological simplicity and to the associated explanatory virtue of unification. Under the guidance of these latter virtues, similarities rather than differences among the phenomena would be stressed.

Mutuality or reciprocity of interaction, sometimes more generally complexity of interaction, is something of a processual companion to the virtue of ontological heterogeneity. While heterogeneity of ontology concerns the existence of different kinds of thing, complexity, mutuality, and reciprocity characterize their interactions. Feminists endorsing this virtue express a preference for theories representing interactions as complex and involving not just joint but also mutual and reciprocal relationships among factors in a process. They explicitly reject theories or explanatory models that attempt to identify one causal factor in a process, whether that be a dominant animal or a "master molecule" like DNA. The work of geneticist Barbara McClintock, as made popular by Evelyn Keller (1983), is often invoked as a model by feminists advocating heterogeneity and complexity. Attentive to individual differences in the maize samples she studied, she also represented causal relations as involving complex interaction. Many feminists are drawn to Developmental Systems Theory (Oyama 2000) by its similar virtues.

Finally, many feminists also endorse the idea that science should be "for the people," that research that promises to alleviate human needs—especially those traditionally attended by women, such as care of the young, weak, and infirm, or feeding the hungry—should be preferred over research for military purposes or for knowledge's sake. While not rejecting curiosity altogether as an appropriate motive of research, these feminists place a greater emphasis on the pragmatic dimension of knowledge, but only in connection with the final virtue in this collection, decentralization of power. Thus forms of knowledge and its application in technologies that empower beneficiaries are preferred to those that produce or reproduce dependence relations. So medical research directed to preventive measures, or to low-cost, easily administered (or self-administered) medications is preferred to research on high-tech, high-maintenance measures. And agricultural research that will assist and empower small farmers is preferred to that which assists capital-intensive agribusiness (see Sen and Grown 1987). Both the feminist pragmatic virtues and their traditional contraries, fruitfulness and refutability, have to do with the expansion of a theoretical approach in an empirical direction. But the relevance of the empirical in the traditional view is within a self-enclosed research context. Applicability and empowerment, by contrast, are directed to the social and practical milieu outside the research context.

Feminist and Traditional Cognitive Values

One might ask why the virtues I've just sketched should be given equal status with the more traditional epistemic virtues with which they contrast. But this question begs another—what is the status of the traditional epistemic virtues?[3] While these are quite frequently invoked as factors closing the gap between evidence and hypotheses revealed by underdetermination arguments, it is not at all evident that they are capable of discriminating between the more and less probable, let alone between the true and the false. Consistency with theories in other domains, for example, only has epistemic value if we suppose these other theories to be true. While they presumably are empirically adequate, additional considerations in favor of their truth will have to consist of other assumptions or theoretical virtues. The probative value of consistency, then, is relative to the truth of the theories with which consistency is recommended.

Simplicity and explanatory power fare no better. While there is an understandable preference for simpler theories when contrasted with theories or models loaded with entities and processes and relationships that do not add to the predictive capacities of the theory, it is not clear that simplicity generally can carry epistemic weight. As is well known, simplicity can be interpreted in different ways. The interpretation contrasting with the alternative virtue of heterogeneity is ontological—the fewer kinds of entities the better, or no more entities than are required to explain the phenomena. As a caution of prudence this has much to recommend it, and it may in some contexts be a useful heuristic. But treating simplicity as an epistemic standard encounters at least three problems:

1. This formulation begs the question of what counts as an adequate explanation. Is an adequate explanation an account sufficient to generate predictions, or an account of underlying processes, and, even if explanation is just retrospective prediction, then must it be successful at individual or at population levels? Either the meaning of simplicity will be relative to one's account of explanation, thus undermining the capacity of simplicity to function as an independent epistemic value, or the insistence on simplicity will dictate what gets explained and how. Feminists have criticized the representation of economics as a certain kind of approach, one that explains phenomena as outcomes of self-interested individuals making choices that maximize their utilities, partly because this definition limits what economics thinks it must explain (as well as, of course, limiting the kind of explanation that will be given).

2. We have no a priori reason to think the universe simple, that is, composed of very few kinds of thing (as few as the kinds of elementary particles, for example) rather than of many different kinds of thing. Or—as Kant taught us—we can give a priori arguments for both theses, nullifying the probative significance of each. There is no empirical evidence for such a view, nor could there be.

3. The degree of simplicity or variety in one's theoretical ontology may be dependent on the degree of variety one admits into one's description of the phenomena. If one imposes uniformity on the data by rejecting anomalies or differences, then one is making a choice for a certain kind of account. If the

view that the boundaries of our descriptive categories are conventional is correct, then there is no epistemological fault in this, but neither is there virtue.

Explanatory power and generality also lose their epistemic allure under close examination. Indeed the greater the explanatory power and generality of a theory, that is, the greater the variety of phenomena brought under its explanatory umbrella, the less likely it is to be (literally) true. Its explanatory strength is purchased at the cost of truth, which lies in the details and may be captured through the filling in of an indefinite series of ceteris paribus clauses (Cartwright 1983). Explanatory power and generality may constitute good reasons for accepting a model or theory if one places value on unifying theoretical frameworks, but this is a value distinct from truth, and it has to be defended on other grounds. Mutuality or reciprocity of influence in an explanatory model is less likely to be generalizable than a linear or unicausal model that permits the incorporation of the explanation of an effect into an explanation of its cause. The explanations of multiple interacting causal factors branch out rather than coalescing. Rather than a vertically ordered hierarchy culminating in a master theory or master science, scientific knowledge would consist of a horizontally ordered network of models.

The feminist and traditional virtues are on a par, epistemologically. Both have heuristic but not probative power. As heuristics, they help an investigator identify pattern or order in the empirical world. They are often transmitted as part of an investigator's training, as part of the common, taken-for-granted background. If we accept that there can be multiple sets of heuristics, pointing in different directions, then we must able to offer some reason for choosing or relying on one set rather than another. I've just argued that greater likelihood of selecting the true cannot be the justification or rationale for relying on the traditional virtues, or what we might call research heuristics. Thus achievement of other goals must constitute the rationale. Some examples will make this clearer.

In particular research contexts, the contrasting heuristics favor different theories, and in some of these contexts this differential favoring has different political consequences. Consider medical research. Uniformity of subjects allows easy identification of effectiveness of pharmaceuticals, but permits no knowledge of their effectiveness for those who are different from the type of subject chosen. In the United States, this meant that, until research was ordered by the director of the National Institute of Health and an act of Congress in the early 1990s, almost nothing was known about the effectiveness or appropriate dosages of medications for women of any race or for nonwhite men. To the extent that the simplicity heuristic is invoked in defense of the previous practice, simplicity has a political valence. One might argue that the situation described was an empirical failure, but it only appears as such in a context in which the lives and well-being of women of any race and nonwhite men are regarded as being as important as those of white men. Empirical adequacy calls for agreement with the data but does not specify which data or even what counts as *all* the data. Simplicity can be used to justify emphasizing similarities over differences, treating anomalous or even systematic differences as insignificant.

Feminist theories of the household, such as those advanced by economists Nancy Folbre (Bittman et al. 2003) or Bina Agarwal (1997), treat the household as comprised of individuals with competing and conflicting interests who must negotiate and bargain with each other. Household consumption patterns are understood as the result of interactions among heterogeneous actors. This model makes the independent and often conflicting interests of the different members of households—spouses, as well as children and elderly parents—visible, rather than assuming that household choices reflect the common interests of all members. On the contrary, the "new economics of the family" championed by, for example, Gary Becker, treats household interests as homogeneous and represented by the choices of the "benevolent patriarch," the actor in the world whose choices determine household consumption patterns. This latter model, of course, exhibits the traditional virtue of simplicity, as it assumes a single, uniformly characterized, economic actor—the rational utility maximizer—but it erases from analytical view gender relations within the household. These models clearly have different political and social implications; one conforms to the traditional, if mythical, nuclear family structure that public policy continues to privilege, and the other does not.

Another, infamous example comes from the biology of reproduction. The traditional view of the process called fertilization in sexually reproducing organisms held that of the two cellular parties to fertilization, the egg and the sperm, one, the egg, was passive, and the other, the sperm, was active. Sperm are repeatedly represented as having to struggle to reach and penetrate the ovum. Another representation of this process stresses the active role of the egg in both stabilizing the highly motile sperm and releasing chemicals that permit its passage through the zona pellucida. This model, which exhibits the alternative virtues of heterogeneity and complexity of causal interaction, has been around in some form at least since the 1930s, but it was dismissed by most biologists investigating sexual reproduction. I am told by some sources that no one now engaged in research on sexual reproduction really believes the first story. I have yet to conduct a review of research articles to confirm that this is correct. But polls of my students indicate that the first story is still featured in many biology textbooks and in teaching. While taking the story seriously in research would continue to privilege an account that replicates and indirectly validates social stereotypes of male activity and female passivity, its relegation to textbooks is not much less insidious. It helps to select into the study of biology those who are entranced by the drama of the heroic sperm, and to select out those left indifferent or convinced of their ancillary role in nature as in society (Martin 1991).

Finally, what Kuhn (1977) called fruitfulness and the feminist pragmatic virtues are not really contraries in their epistemic relevance, since both effectively call for empirical consequences, but they have different pragmatic valences. The fruitfulness of a theory is its ability to generate problems for research. This does not argue for the truth of a theory but for its tractability, that is, for its capacity to have empirical data brought to bear on it. Fruitfulness so understood may be less an intrinsic feature of a theory or model than a matter of the material and intellectual instruments available for producing relevant data, as well as other theoretical and empirical developments in associated fields that make articulation of the theory possible. A model or theory or concept may be fruitless in one century and fruitful in another. (Think of heliocentrism for Aristarchus and for Copernicus, or relativity for Leibniz and for Einstein.) The feminist pragmatic virtues instead seek the importance of empirical consequences in certain areas: in the world of human life as well as in the laboratory. And, as noted, the most explicitly political of the feminist virtues requires in addition that the mode of applicability involve empowerment of the many rather than the concentration of power among the few.

Some thinkers about the sciences have rejected altogether the distinction between pure and applied science that lies behind the standard treatment of refutability or fruitfulness as a virtue, that is, as a criterion of theory evaluation or selection. Contemporary science, on this view, is better understood as technoscience, inquiry into nature that is inseparable from its technological infrastructures and outcomes (Latour 1987). Within this framework, the feminist pragmatic virtues could be understood not as a rejection of "pure science" but as a recognition of the technologically driven nature of science and a call for certain technological infrastructures and outcomes over others. Rejecting the conventional distinction between pure and applied science facilitates the rejection of the idea that scientists bear no responsibility for how their work is used. Whether one takes on the pure/applied distinction or not, the feminist pragmatic virtues can be a vehicle for bringing considerations of social responsibility back into the center of scientific inquiry.

While all of these points could be further developed, I have, for each of the more mainstream epistemic values, indicated why their epistemic status is no greater than that of the alternatives advocated by feminist researchers and philosophers. Neither is more conducive to truth than the other. Instead they can be understood as heuristics valuable for guiding investigation to produce knowledge of the sort that is required in a given context of inquiry. Both sets of heuristics have political valence in particular contexts of application (the examples from medicine, economics, and cell biology could be multiplied). In general, however, only the upstart or oppositional research community will acknowledge the relation of its research heuristics to its social and political values and aims.

One might well ask of the alternative virtues/heuristics I have described what makes them feminist. Many answers have been offered, but I think this is the wrong question. They are, after all, not advocated exclusively by feminists, but also by other oppositional scientists, for example, the dialectical biologist imagined by Richard Levins and Richard Lewontin (1985). They characterize work by scientists who would reject the label "oppositional" and serve as alternatives for a scientific community larger (or different) than the feminist one.

The question ought instead to be what recommends the alternative heuristics to feminists? As I have suggested elsewhere, what *ought* to recommend them to feminists is that they (do or could) serve feminist cognitive goals. What makes feminists feminist is the desire to dismantle (eliminate, end) the oppression and subordination of women. This requires identification of the mechanisms and institutions of female oppression and subordination, that is, the mechanisms and institutions of gender. The cognitive goal of feminist researchers therefore, is to reveal the operation of gender by making visible both the activities of those gendered female and the processes whereby they are made invisible, and by identifying the symbolic and institutional mechanisms whereby female gendered agents are subordinated. What ought to recommend these virtues to feminists, then, is that inquiry guided by these heuristics and theories characterized by these virtues are more likely to reveal gender than inquiry guided by the mainstream virtues. The examples I just outlined are instances of the relation between possession of these virtues and the revealing or concealing of gender. Feminists deem problematic research that naturalizes or disguises dominance relations. Feminist analysis demonstrates of certain relations that they are relations of dominance. Resistance to concealment of dominance is the feminist basis of resistance to reductionism. There is undoubtedly more to be said here as well, including consideration of other possible theoretical virtues, other cognitive aims, and the relations of these virtues to other (noncognitive) values endorsed by feminists and to values endorsed by other communities of inquiry. What I have done so far is to propose a kind of pragmatic or teleological structure. What are some of the consequences of taking this structure seriously?

Epistemological Reflections

What can these virtues and their relations tell us about the prospects for a feminist or nonreductionist or alternative science practice based on them? First of all, although the virtues have been endorsed by feminists (although not by all feminists) and can be discerned at work in feminist appraisal, their subordination to a broader cognitive goal means that they are not in and of themselves feminist theoretical virtues, or to put it another way, such subordination means that these alternative virtues will not necessarily be a part of a feminist epistemological kit. They have no intrinsic standing as feminist theoretical virtues or virtues for feminists, but only a provisional one. For as long as and to the extent that their regulative or heuristic role can promote the goal of revealing gender, and as long as revealing gender remains the primary goal of feminist inquiry, they can serve as norms or standards or guides in feminist inquiry. It is possible, however, that in different contexts they would not promote feminist cognitive goals, or that those goals themselves might change in such a way that other cognitively regulative norms would be called for. Indeed, to the extent that feminists dissent from the virtues, they may either be urging a change of feminist cognitive goals or claiming that the goals are not served by the virtues discussed here. There could be multiple sets of feminist cognitive virtues corresponding to different conceptions of what feminist cognitive goals are or should be. The concept of gender has itself changed as a consequence of feminist inquiry. Recognizing the disunity both of gender and of forms of gender subordination might require either a change in cognitive aim or a change in the virtues. Not only that, mainstream science may overtake or co-opt themes in outsider science: many sciences are now embracing heterogeneity and complexity (for example, network thinking in sociology, or recent work on RNA transcription in molecular biology), but they may do so in ways that again render gender invisible. If this is so, feminists will need heuristics other than those they advocated in the heyday of reductionist science. This suggests the picture of a pool of heuristics, neutral in themselves, which may serve different social values in different intellectual and political situations.

Secondly, the normative claim of these values/virtues/heuristics is limited to the community sharing the primary goal. On those who do not share it they have no claim. To expand this point, the alternative values are only binding in those communities sharing a cognitive goal that is advanced by those values. Their normative reach is thus local. In emphasizing the provisionality and locality of alternative virtues, this account contrasts quite sharply with accounts offered or implied by advocates of the traditional virtues which, whether advanced as (purely) epistemic, or as Kuhn proposed, constitutive of science, are represented as universally binding. I've said that arguments for the alternative virtues must appeal to and argue on behalf of the cognitive goals they are thought to serve. What does application of this pragmatic-teleological structure to the traditional virtues show? What is missing from most advocacy on their behalf is the articulation of a cognitive goal that would ground them, or that they would serve. If there are multiple possible sets of heuristics or virtues, then no set is self-announcing, and the structure of their justification must be the same as that for the particular set of alternative heuristics I discussed. The cognitive goal to be attained by reliance on the traditional set has yet to be identified. (For reasons made clearer below, truth won't do.) Any set, then will be only provisional and locally binding.

There is a further consequence of treating these as heuristics.[4] Typically, the discussion of values or virtues has concerned considerations to be raised in the face of two equally well-developed and equivalently empirically adequate theories. Heuristics come into play earlier in research, when one is trying to formulate models or make choices among directions to pursue. The supposition that there was only one set of scientific values permitted the traditional values to play a role in guiding the development of models as well as a role in theory decision. This may in fact be their most pernicious effect—that of truncating investigation of alternatives because they don't exemplify the traditional virtues. I have argued that the traditional and alternative sets are on a par, epistemically speaking. In light of the foregoing remarks, this means I would urge treating the members of both as heuristics, employed in the context of theory and model development, rather than assigning them probative status operative at the end of inquiry. They do not function as arbitrators to be appealed to when faced with an underdetermination situation. As many philosophers have argued, scientists themselves rarely face the sorts of decisions represented in idealized underdetermination situations. This, I would argue, is because the heuristics, as well as other assumptions that close the underdetermination gap, have played a role all along: they have shaped questions, guided the selection and representation of data and the choice of methods, and granted prima facie plausibility to certain models and hypotheses over others.

To bring these thoughts to bear on the politics of knowledge, let me place all this within the overall critical contextual empiricism I advocate.[5] Data (measurements, observations, experimental results) acquire evidential relevance for hypotheses only in the context of background assumptions. These acquire stability and legitimacy through surviving criticism. Justificatory practices must therefore include not only the testing of hypotheses against data, but the subjection of background assumptions (and reasoning and data) to criticism from a variety of perspectives. Thus, intersubjective discursive interaction is added to interaction with the material world under investigation as components of methodology. From a normative point of view, this means articulating conditions for effective criticism, typically specifying structural features of a discursive community that ensure the effectiveness of the critical discourse taking place within it. I have suggested four such conditions: the provision of venues for the articulation of criticism; uptake (rather than mere toleration) of criticism; public standards to which discursive interactions are referenced; and equality (or tempered equality) of intellectual authority for all members of the community.

The public standards that regulate the discursive and material interactions of a community are both provisional and subordinated to the overall goal of inquiry for a community. Truth simpliciter cannot be such a goal, since it is not sufficient to direct inquiry. Rather, communities seek particular kinds of truths. They seek representations, explanations, technological recipes, and so on. Researchers in biological communities seek truths about the development of individual organisms, about the history of lineages, about the physiological functioning of organisms, about the mechanics of parts of organisms, about molecular interactions, and so on. Research in other areas is similarly organized around specific questions. Which kinds of truths are sought in any particular research project is determined by the kinds of questions researchers are asking and the purposes for which they ask them, that is, the uses to which the answers will be put. Different sets of heuristics (consisting of rules of data collection, including standards of relevance and precision; inference principles; and the epistemic or cognitive values) will satisfy the different cognitive goals. Truth is not opposed to social values, indeed it *is* a social value in the sense that the demand that scientific inquiry provide truths rather than falsehoods is a social demand, but the regulatory function of truth is directed or mediated by other social values operative in the research context.[6]

One consequence of taking this epistemological view is pluralism. Other philosophers have advanced pluralism as a view about the world, that is, as the consequence of a natural complexity so deep that no single theory or model can fully capture all the causal interactions involved in any given process. While this may be the case, the epistemological position I am advocating is merely open to pluralism in that it does not presuppose monism. It can be appropriate to speak of knowledge even when there are ways of knowing a phenomenon that cannot be simultaneously embraced. Whether or not it is appropriate in any given case depends on satisfaction of the social conditions of knowledge mentioned above. When these are satisfied, reliance on any particular set of assumptions must be defended in relation to the cognitive aims of the research. These are not just a matter of the individual motivations of the researchers but of the goals and interests of the communities that support and sustain the re-search. On the social view, all of these must be publicly sustained by surviving critical scrutiny. Thus, social values come to play an ineradicable role in certain contexts of scientific judgment.

I've already indicated why the feminist virtues or any set of alternative theoretical virtues could not be superseded by the traditional virtues. Two fur-ther objections must be addressed. One might ask whether there is not a set of cognitive values different from both the heuristics identified as traditional and those identified as alternative, which would constitute universal norms. Per-haps the verdicts of provisionality and partiality are the consequence of looking at the wrong values. But this objection must provide examples of values that could be universally binding. The only characteristics of theories or hypotheses that might qualify are truth or empirical adequacy. But truth in the context of theory adjudication reduces to empirical adequacy—truth of the observational statements of a theory. And empirical adequacy is not sufficient to eliminate all but one of a set of contesting theories. It is because the purely epistemic is not rich enough to guide inquiry and theory appraisal that the heuristics discussed earlier come into play. (For arguments about the insufficiency of truth simpliciter, see Anderson 1995 and Grandy 1987.) One might, alternatively, specify qualities of inquirers that count as virtues, for example, open-mindedness and sensory or logical acuity, but these are not theoretical, but personal virtues, not public standards of critical discourse but qualities required to participate constructively in such discourse.

Secondly, one might resist the identification of competing sets of virtues and suggest the integration of the two sets of virtues. There are two difficulties with this suggestion. In particular contexts of inquiry, virtues from the two sets recommend nonreconcilable theories (see above, and Longino 1996). Moreover, integration can be understood in at least two ways, each involving quite different presuppositions. It might be proposed as fulfilling a commitment to unified science, but that commitment needs support. It might, however, be proposed as a way of realizing theoretical pluralism within a single community. This presupposes the value of the (particular) diversity of models that inclusion of both sets of values in a community's standards might produce. If so, what is called for is not integration of the virtues by one research community, but the tolerance of and interaction with research guided by different theoretical virtues, the construction of larger meta-communities characterized by mutual respect for divergent points of view, that is, by pluralism.[7]

Politics of Knowledge

Within this scheme, then, the traditional and alternative heuristics or virtues constitute partially overlapping but distinctive sets of public community standards. That is, they serve both to guide the development of models and hypotheses relevant to the empirical situation under investigation and, when generally accepted, to regulate discourse in their respective communities. They are not fixed but can be criticized or challenged relative to the cognitive aims they are taken to advance or to other values assigned higher priority and they can in turn serve as grounds for critique. Nor is criticism limited to intracommunity discourse. The areas of overlap or intersection make possible critical interaction among as well as within communities. Generalizing from what I have argued earlier, the public standards that I argue must be a component feature of an objective or reliable scientific community will be binding only on those who share the overall cognitive goal that grounds those standards, and who agree that the standards do indeed advance that cognitive goal. Such agreement must itself be the outcome of critical discursive interactions in a context satisfying conditions of effective criticism. As the virtues understood as public standards are subordinated to the advancement of a specific cognitive aim that may change, they must be understood as provisional. As they are binding only on those who share that aim, they must be understood as partial.

This way of thinking about knowledge and inquiry involves a shift in attention away from the outcomes or products of inquiry, whether these are theories or beliefs, to the processes or dynamics of knowledge production. The ideal state, from an epistemological point of view, is not having the single best account, but the existence of a plurality of theoretical orientations that both make possible the elaboration of particular models of the phenomenal world and serve as resources for criticism of each other. Pragmatically, of course, selection must be made of a model that will guide action, but if we arbitrarily limit those in contention by arbitrary exclusion of alternative heuristics, we risk under- or ill-informed action or policy.

However, developing a model or hypothesis sufficiently that it can contribute to critical interaction and be applicable to empirical problems requires resources—time, intellectual space, material resources.

This is where politics come in. If we propose that models of natural processes developed within an approach that holds a given set of virtues paramount are part of a plurality of adequate representations answering to different cognitive aims, we relieve the feminist, or indeed any scientist, of the burden of completeness or finality. We come to see knowledge as both dynamic and partial. This plurality is not merely the existence of alternative models and differently constituted scientific communities. When religious minorities struggled for pluralism, they struggled for tolerance. The pluralism I advocate for philosophy of science, and by extension, science, does not require mere toleration. Scientific pluralism involves interaction among different approaches—a mutual taking seriously, or "uptake." Not every crackpot idea needs to be accorded the same seriousness, but acknowledgement of plurality requires much greater care in dismissing alternative perspectives. The condition of equality of intellectual authority draws attention to the present unequal distribution of intellectual authority. Equality of intellectual authority does not come into being because a philosophical argument contends it is a necessary condition of genuine or fully reliable knowledge production. It must be fought for in the following ways:

- By contesting the practices of marginalization that make members of certain social categories—women or members of ethnic minorities—

invisible, even when their contributions to a given undertaking are equal to or greater than those of male or white colleagues. (To those who think such marginalization is no longer a problem, one might note, for example, citation patterns in publication, or to whom ideas raised in a meeting are attributed; see Fricker 2003.)

- By attending to the material conditions that give some voices and perspectives greater authority than others and working to change them.

- By attending to the material and social consequences of adopting a particular model or representation of a given process and actively seeking alternatives (and the investigative tools required to produce them) when necessary.

- By being vigilant to possibilities of co-optation (for example, the virtues of heterogeneity and complexity, when detached from a cognitive interest in exposing dominance relations, may be used in ways that reinforce inequality).

These different avenues of action are each important. The critics of the sciences might be right about the content and methods of current sciences, but being right is not enough. To effectively challenge current and invidious presuppositions, feminist inquirers must join with others marginalized by current structures of power and interest to claim and create our own spaces for the production of scientific knowledge—knowledge that does not naturalize relations of domination but that offers other ways to interact with the natural world, and by extension with one other. We must also find ways to communicate these alternatives in order that science's public as well as the "science establishment" take them seriously. Better knowledge alone will not change the social world, partly because the social world itself must change in order that other knowledge emerge. However, in a social world so dependent on knowledge and on science, we can't afford to change only science or only the world, but must continually work to change both.

NOTES

1. For advocates, see Kuhn (1977) and McMullin (1983); for critics, van Fraassen (1980).

2. The following paragraphs draw on previously published material; see Longino (1996).

3. The following paragraphs draw on previously published material; see Longino (1997).

4. I am grateful to participants in the Notre Dame Bielefeld Conference on Science and Values for pressing this point upon me.

5. Its most recent expression is in Longino (2002).

6. For a more extensive discussion of truth (or semantic success) of theoretical claims, see Longino (2002).

7. Of course, one might object that what results is not science at all, properly understood, but a failed attempt at science. If by "science" one means some idealized rational practice, perhaps so. But if by "science" one means the attempt to describe and understand the natural and social worlds by the kind of limited cognitive agents humans are, then such pluralism is inevitable.

REFERENCES

Agarwal, Bina. 1997. "'Bargaining' and Gender Relations: Within and Beyond the Household." *Feminist Economics* 3 (1): 1–51.

Altmann, Jeanne. 1974. "Observational Study of Behavior: Sampling Methods." *Behavior* 49:227–67.

Anderson, Elizabeth. 1995. "Knowledge, Human Interests, and Objectivity in Feminist Epistemology." *Philosophical Topics* 23:59–94.

Bittman, Michael, et al. 2003. "When Does Gender Trump Money? Bargaining and Time in Household Work." *American Journal of Sociology* 109 (1): 186–214.

Cartwright, Nancy. 1983. *How the Laws of Physics Lie*. Oxford: Oxford University Press.

Fricker, Miranda. 2003. "Epistemic Injustice and a Role for Virtue in the Politics of Knowing." *Metaphilosophy* 34:154–73.

Grandy, Richard. 1987. "Information Based Epistemology, Ecological Epistemology, and Epistemology Naturalized." *Synthese* 70 (1): 191–203.

Haraway, Donna. 1986. "Primatology Is Politics by Other Means." In *Feminist Approaches to Science*, ed. Ruth Bleier, 77–118. Elmsford, NY: Pergamon Press.

Hrdy, Sarah Blaffer. 1981. *The Woman that Never Evolved*. Cambridge, MA: Harvard University Press.

Keller, Evelyn. 1983. *A Feeling for the Organism*. San Francisco: W. H. Freeman and Co.

———. 1995. *Refiguring Life*. New York: Columbia University Press.

Kuhn, Thomas. 1977. "Values, Objectivity and Theory Choice." In *The Essential Tension*, 120–39. Chicago: University of Chicago Press.

Latour, Bruno. 1987. *Science in Action*. Cambridge, MA: Harvard University Press.

Levins, Richard, and Richard Lewontin. 1985. *The Dialectical Biologist*. Cambridge, MA: Harvard University Press.

Longino, Helen E. 1996. "Cognitive and Non-cognitive Values in Science." In *Feminism, Science, and the Philosophy of Science*, ed. Lynn Nelson and Jack Nelson, 39–58. London: Kluwer Publishers.

———. 1997. "Feminist Epistemology as a Local Epistemology." *Proceedings of the Aristotelian Society*, Suppl. 71:19–35.

———. 2002. *The Fate of Knowledge*. Princeton: Princeton University Press.

Martin, Emily. 1991. "The Egg and the Sperm." *Signs: Journal of Women in Culture and Society* 16 (3): 485–501.

McMullin, Ernan. 1983. "Values in Science." In *PSA 1982*, vol. 2., ed. Peter D. Asquith and Thomas Nickles. East Lansing, MI: Philosophy of Science Association.

Oyama, Susan. 2000. *The Ontogeny of Information*. Durham, NC: Duke University Press.

Quine, Willard V. O., and Joseph Ullian. 1978. *The Web of Belief*. New York: Random House.

Sen, Gita, and Caren Grown. 1987. *Development, Crises and Alternative Visions: Third World Women's Perspectives*. NY: Monthly Review Press.

Sperling, Susan. 1991. "Baboons with Briefcases: Feminism, Functionalism and Sociobiology in the Evolution of Primate Gender." *Signs: Journal of Women in Culture and Society* 17 (1): 1–27.

van Fraassen, Bas. 1980. *The Scientific Image*. Oxford: Oxford University Press.

Zihlmann, Adrienne. 1978. "Women in Evolution, Pt. II." *Signs: Journal of Women in Culture and Society* 4 (1): 4–20.1.

4

REPLACING THE IDEAL OF
VALUE-FREE SCIENCE

JANET A. KOURANY
University of Notre Dame

THE IDEAL OF VALUE-FREE SCIENCE HAS ENJOYED a long and distinguished career. Some see it already flourishing in ancient times with the Platonic separation of the theoretical and the practical and the privileging of the theoretical. Most, however, see it emerging with the rise of modern science in the seventeenth century and the idea that nature is merely matter in motion, devoid of qualities such as good and evil. They see it as well in the seventeenth-century idea that the study of nature is distorted by ethical concerns in much the way Bacon claimed such study is distorted by the various idols he described. The ideal of value-free science is seen flourishing again in the eighteenth century with Hume's separation of "ought" from "is," and in the nineteenth century with the push toward academic specialization and the emphasis on the increasingly technical specialties and subspecialties of science as impartial resources for the solution of social problems. And the ideal of value-free science is seen flourishing once again in the twentieth century, with the many historical and philosophical and sometimes even sociological accounts of science in which social values either play no role at all or at least no very helpful role (for more details of this history, see Proctor 1991).

But all that is past. The ideal of value-free science, many now say, has finally retired from the scene, largely due to advice provided by the history, sociology, and philosophy of science. Historical scholarship, for example, has suggested that the work of even the greatest scientists—even scientists like Boyle, Darwin, and Freud, and even, perhaps, the great Newton and Einstein themselves—was shaped by social values (see, for example, Bernal 1971; Merchant 1980; Elkana 1982; Shapin and Schaffer 1985; Gilman 1993; Ruse 1999; Potter 2001). If our conception of science, including our conception of objective science, is to be true to actual science, it can hardly ignore such science as this. Sociological research, in addition, has suggested that such value-informed science is all but inevitable. Indeed, any scientific contribution, we have been told, is a product of a particular time and place, of a particular social and cultural location, of particular interests and values; a "view from nowhere," from a psychological and sociological vantage point, is simply naive (see, for example, Knorr-Cetina 1981; Knorr-Cetina and Mulkay 1983; Latour 1987). The ideal of value-free science, in short, seems unlikely ever to be fulfilled—at least seems unlikely to be a viable ideal, useful for actual science. Philosophical analysis, finally, has gone one step further. It has challenged the very distinction between social values and the scientific—the distinction between, for example, social values and economists' data about poverty, or sociologists' and psychologists' measures of domestic abuse, or archaeologists' accounts of human evolution and human flourishing, or medical researchers' criteria of health and disease (see, for example, Putnam 2002; Dupré 2007). The ideal of value-free science, in short, according to this line of reasoning, may ultimately be incoherent. The history, sociology, and philosophy of science, then, have done much to bring about the retirement of the ideal of value-free science. What they have not made especially clear is what should now take its place. This is the task of the present investigation.

The Old Ideal's Job Description

Well, what was its place, what roles did the ideal of value-free science play, at least in recent times? Consider, for example, the interdisciplinary area of feminist science studies, one of the main instigators of the retirement of the ideal of value-free science. Feminists who do science or who reflect on science, including feminist historians, sociologists, and philosophers of science, have been engaged for years with the question of social values in science. They have exposed sexist and androcentric values operating in such fields as medical research, biology, psychology, sociology, anthropology, economics, political science, and sometimes even physics and chemistry (see, for some examples, Keller 1985, 1992; Hubbard 1990; di Leonardo 1991; Fausto-Sterling 1992; Kramarae and Spender 1992; Rosser 1994; Spanier 1995; and Nelson 1996a, 1996b). And they have exposed other sorts of values operating in science as well—heterosexist values and racist values and capitalist values, for example (see Haraway 1989 and Harding 1993, 1998). What's more, feminist scientists have allowed feminist values to shape important aspects of their research—from research questions and assumptions to concepts and hypotheses and even methods of data collection and modes of theory evaluation. And feminist philosophers and historians of science have cheered these feminist scientists on (see, for example, Schiebinger 1999 and Creager, Lunbeck, and Schiebinger 2001, as well as the work of Helen Longino, especially her 1990 and her classic 1987).

A politically hopeful way to understand this scene was originally provided by the ideal of value-free science.[1] Value-free science, remember, was an *ideal*, not a straightforward description of science. It specified what science ought to be like if it were to serve up genuine knowledge. The ideal of value-free science, therefore, was not obviously challenged by feminists' exposure of androcentric and sexist values in science as well as heterosexist and racist values. On the contrary, the ideal of value-free science provided a rationale for what took place—for the way feminist scientists judged sexist and racist science to be bad science and the ways they sought to rid science of such sexism and racism. Indeed, many feminist scientists who exposed sexist and racist values in science sought traditional scientific remedies. They took to task mainstream scientists for failing to abide by accepted standards of concept formation and experimental design and interpretation of data and the like (see, for example, Bleier 1984; Hubbard 1990; Fausto-Sterling 1992). If only such standards were rigorously followed, they suggested, the problem of sexism and racism in science would be, at the very least, much reduced. Other feminist scientists explored new ways of screening out the offending values once and for all—new methodologies that would reform the science (e.g., Eichler 1980, 1988) or new pedagogies that would reform the scientists (e.g., Rosser 1986, 1990, 1995, and 1997). Even some of the scientists who consciously shaped their research in accordance with feminist values pursued such approaches. For example, they treated the function of feminist values in their research as purely motivational and not really a part of that research. Or they treated feminist values not as an alternative to sexist values but as a new kind of methodological control to prevent the entry of sexist values into their research. "We have come to look at feminist critique as we would any other experimental control," one widely quoted group of scientists said. "Feminist critique asks if there may be some assumptions that we haven't checked concerning gender bias. In this way feminist critique should be part of normative science. Like any control, it seeks to provide critical rigor, and to ignore this critique is to ignore a possible source of error" (The Biology and Gender Study Group 1988, 61–62; cf. Eichler 1980, 118: "Feminist science is non-sexist" science). For all these scientists, then, the ideal of value-free science could have, and sometimes did, provide an explicit rationale for their various responses to both feminist and sexist science. At the same time, the ideal of value-free science provided hope that all could be made right—that science would be able, finally, to provide objective information about women and, in the process, expose and remove society's prejudice against women, not simply reinforce and perpetuate it.

In feminist science studies, then, the ideal of value-free science played both an epistemic role and a political role—suggested both a way to achieve objective knowledge and a way to achieve social reform. And, of course, the two roles were connected, for the epistemic role was to make possible the political role, objective knowledge was to make possible social reform. With the retirement of the ideal of value-free science, what new ideal can now play these roles?

Helen Longino's Candidate for the Job

An especially promising candidate—and one groomed with the challenges of the feminist science studies terrain clearly in view—is a candidate put forward by Helen Longino (2002; see also 1990). According to Longino, no scientific method, however rigorous and however rigorously applied, can be guaranteed to screen out the various values and interests that scientists from their different social locations bring to their research. To be sure, scientists' values and interests can and do determine which questions they investigate and which they ignore, can and do motivate the background assumptions they accept and those they reject, can and do influence the observational or experimental data they select to study and the way they interpret those data, and so forth. The ideal appropriate for science, then, is not the ideal of value-free science, Longino argues, but the "social value management" ideal of science (2002, 50). According to this ideal, all social values should be welcomed into science—indeed, encouraged—and all social values, and the science they engender, should be subjected to criticism. So there is a kind of neutrality here, akin to the old ideal of value-free science. The only restrictions, in fact, have to do with the social organization of scientific communities. These communities, Longino insists, will have to have first, public venues for criticism; second, publicly recognized standards by reference to which such criticism can be made; third, "uptake" of such criticism (that is, the criticism will have to be taken seriously and responded to); and fourth, "tempered equality" of intellectual authority among all parties to the debate, among whom "all relevant perspectives are represented" (2002, 131). The output of a scientific community can constitute "knowledge," in short, even if that output is inspired and informed by social values, if the community meets these four conditions and the output conforms sufficiently to its objects to enable the members of the community to carry out their projects with respect to those objects.

So much for preliminaries. What credentials does Longino's candidate offer for the position now vacated by the ideal of value-free science? In her recent book, *The Fate of Knowledge*, Longino does much to exhibit these credentials. She shows her candidate informed by some of the most important findings of historical/sociological research. She shows her candidate informed, as well, by the enduring insights of epistemological reflection. At the same time, she shows her candidate able to integrate all these findings and insights into a coherent account of science, an account free of the confusions that have frequently accompanied them. Longino thereby suggests that her candidate is unsullied by the historical, sociological, and philosophical disclosures that brought about the retirement of the old ideal of value-free science. Perhaps most important, however, Longino implies that her candidate exemplifies just what we have been looking for to replace that ideal, just what we mean by terms such as "knowledge" and "objectivity." Granted, this last accolade somewhat strains credulity. After all, all of us, before Longino wrote, thought we had some handle on the meaning of these terms, but doubtless most of us had no handle on what Longino describes—tempered equality and public venues and uptake and the rest—not even a preanalytic, prearticulated version of what she describes.[2] No matter. If Longino's candidate is most comfortable in the dress of traditional analytic philosophy, then it is in the dress of traditional analytic philosophy that we shall conduct its interview.

The Interview of Longino's Candidate

On, then, to the imaginary tale, de rigueur in traditional epistemology. But this time the tale is not about some solitary epistemic agent named Smith, his lucky happenstances, and his (usually unsuccessful) claims to knowledge, as in days of old, but about a scientific *community* named Smith—or rather, named PE-TERS, that is, the Privileged, Exclusive, Talented, Elite, Royal Society. PETERS is made up of a subset of the privileged and talented of society *S* but PETERS is also a very elite society, very exclusive. It excludes all those, albeit sometimes talented persons, who fall into various unfavored classes (the non-privileged, the underprivileged). And PETERS has power—it is, after all, a royal society. So PETERS, knowing where its bread is buttered and also sharing in the perspectives of the butterers, pursues a particular kind of cognitive enterprise, one that serves its particular needs and interests. PETERS, of course, is a scientific society, concerned with understanding the world and interacting with it successfully. But PETERS is also a privileged, exclusive, talented, elite, royal scientific society, and that leaves a definite mark on the parts of the world it seeks to understand and the ways it seeks to interact with them. So, for example, PETERS investigates physical and chemical questions related to its concern with war-making and military preeminence, PETERS investigates biological and psychological questions related to its concern with the maladies that afflict the privileged and the reasons they are superior nonetheless, PETERS investigates archaeological questions related to its concern with the routes by which the privileged have achieved their superior state of development, and so forth. And PETERS's concepts and theories and models and methods and standards and values reflect these concerns, these privilegecentric and privilegist goals.

Our question is: Would PETERS, over time, produce *knowledge* for its members? We can imagine that PETERS regularly holds conferences and publishes journals in which all its members are encouraged to participate and in which all are treated equally. We can imagine that in these venues prolonged and frequently heated critical exchanges take place, exchanges that pay scrupulous attention to shared standards. We can imagine that follow-up exchanges regularly take place, as well. And we can imagine that the intellectual products that emerge from all of this activity conform well enough to their subject matters to enable PETERS to pursue its (privilegecentric and privilegist) projects to its (or rather, its members') satisfaction. We can even imagine that, after some time, PETERS invites, even encourages, members of the underprivileged classes—at least their talented members—to join its ranks, master its methods and standards and values and concepts and models and theories, and contribute to its (privilegecentric and privilegist) projects; we can even imagine that PETERS encourages these underprivileged ones to develop "alternative points of view" that can serve as a "source of criticism and new perspectives" (Longino 2002, 132) so that finally "all relevant perspectives are represented" (Longino 2002, 131) in PETERS's exchanges, that is to say, all perspectives relevant to the satisfaction of PETERS's privilegecentric and privilegist goals. Would PETERS now be producing *knowledge* for its members? It would seem that Longino's candidate must answer "yes," though it should answer "no." And, would that knowledge—if it is knowledge—be free of privilegecentric and privilegist prejudices, and thereby a suitable springboard from which to bring about social reform in society *S* rather than a reinforcement of those same prejudices? It would seem that we must answer "no."

"Stop the interview," I hear you saying. "It's unfair! It's rigged! Longino's candidate is getting pushed in a direction it does not want to go. When Longino says that in order for a scientific community's critical interactions to generate knowledge 'all relevant perspectives' must be represented, she does not only mean all perspectives that might serve that community's goals; she means all perspectives that might relate in any way at all to those goals, that is to say, all perspectives that might support them, or clarify them, or develop them, or add to them, or revise them, or replace them, and so forth. 'Such criticism,' she says, 'may originate from an indeterminate number of points of view, none of which may be excluded from the community's interactions without cognitive impairment'(2002, 133)."

Okay. Start the interview again. Imagine once again that PETERS finally encourages members of the underprivileged classes to join its ranks and develop alternative points of view, *all relevant* alternative points of view, that can serve as sources of criticism and new perspectives. Would PETERS now be producing knowledge for its members? And, would that knowledge, if it is knowledge, be free of privilegecentric and privilegist prejudices, and thereby a suitable springboard from which to bring about social reform in society *S*? We cannot now say simply that PETERS's cognitive output would have to serve its original privilegecentric and privilegist goals and thereby serve the status quo in society *S*, for over time PETERS's cognitive output might evolve in all sorts of ways as a result of the critical discourse occurring in it. The underprivileged ones in PE-TERS, though trained in its privilegecentric and privilegist research traditions, might come to have the wherewithal to develop alternatives to some of those traditions, perhaps aided by changes over time in PETERS or in PETERS's science or in PETERS's surrounding society *S*. The underprivileged ones in PE-

TERS might even succeed over time in building significant support for some of these alternatives, might even succeed in crystallizing new research traditions around some of them that parallel in many ways the older traditions, might even bring about the replacement of some of the older traditions. Women, after all, originally largely excluded from Western science and then, when included, trained in its androcentric and sexist research traditions, still came to have the wherewithal to develop alternatives to some of those traditions, aided by the sheer numbers of women—the "critical mass" of women—in some research areas, and aided by the women's movement in society at large as well as by changes in other academic fields. Women even succeeded over time in building support for some of these alternatives, even succeeded in crystallizing new feminist research programs around some of them to compete with the older programs, even succeeded in replacing some of the older programs. The underprivileged ones in PETERS, though trained in its privilegecentric and privilegist research traditions, then, *might* come to have the wherewithal to replace them. But then again, they might not. They might not have egalitarian political movements in society S to aid them, they might have political backlash instead; they might be stymied by available mathematical resources or instrumental technologies or preferred modes of analysis; they might be affected by funding cutbacks or staffing problems or family needs. Certainly women in science have been thwarted by such factors as these, and certainly women in science have met with far less than unbridled success in trying to rid science of sexism and androcentrism.

So what is the upshot? If what PETERS produces is *knowledge* for its members according to Longino's candidate, this knowledge need not be free of privilegecentric and privilegist prejudices,[3] and it need not be a suitable springboard from which to bring about social reform in society S. If Longino's candidate fulfills the epistemic role of the ideal of value-free science, in short, it still may not fulfill the political role.[4] Well, so what? Had the ideal of value-free science been acceptable, it would have provided a way to rid science of sexist and racist values and the like, and thereby promote social equality. But the ideal of value-free science was *not* acceptable. So why should its successor have actually to do what it, itself, merely promised but could not actually do? Why, in short, should the successor of the ideal of value-free science have to play a political role along with an epistemic role? Then again, if we excuse the successor of the ideal of value-free science from playing its predecessor's political role, we not only lose scientific knowledge as an ally in the fight for social justice, we set scientific knowledge up as part of the problem—part of what reinforces and perpetuates prejudice rather than exposes and removes it. Is there a better way to go?

A Second Candidate Steps Forward

There is another candidate for the position now vacated by the ideal of value-free science—a less sophisticated candidate, by far, than Longino's, but with a certain down-to-earth, homespun charm. It can be called the "ideal of socially responsible science." Rather than strive to exclude all social values from science, as the ideal of value-free science directs scientists to do, or to include all social values in science but subject them all to criticism, as Longino's social value management ideal of science directs scientists to do, the ideal of socially responsible science directs scientists to include only specific social values in science, namely the ones that meet the needs of society. This is the kind of ideal to which many feminist scientists now subscribe, for of course one of the most important needs of society—of both men and women—is justice, and equality between men and women is one aspect of that justice. Thus you find feminist scientists now explicitly shaping their research in accordance with egalitarian social values.

Consider, for example, a new psychological research program described by Carolyn West, concerned with the problem of domestic violence in the United States (West 2002, 2004). The aim of this program is complex: to articulate the similarities in intimate partner violence within the black and white communities of the United States without negating the experiences of black women, and simultaneously to highlight the differences within the black and white communities without perpetuating the stereotype that black Americans are inherently more violent than other ethnic groups. This aim requires charting a new course for research. For example, it requires broadening the definition of partner violence to include psychological, emotional, verbal, and sexual abuse as well as physical abuse. It also requires changing the ways violence is measured—from merely counting violent acts and measuring their severity (which focuses on discrete *male* behaviors) to taking into account the contexts, motives, and outcomes of the violent acts (which focuses on *female* experiences) using a combination of qualitative and quantitative research methods, including listening to the voices of battered women. All this dramatically transforms the picture of racial similarities and differences drawn from past research—the picture according to which, for example, black women, when compared to their white counterparts, are significantly more likely to sustain and inflict aggression, especially aggression involving weapons and culminating in hospitalization. The new research program involves other changes as well: for example, a revision of measurement scales to reflect more than the experiences of white European Americans, taken as the norm; and investigations of within-group differences in the black and white communities to determine whether what appear to be racial differences are not simply socioeconomic differences instead. And the program involves integrating participants into every stage of the research process, from planning to implementing, interpreting, and disseminating results, in order to reduce one-sided research interpretations. The result is the kind of research that both motivates social reform and helps to bring it about.

The ideal of socially responsible science makes research such as West's a model of what science should be like. It applauds funding initiatives that prioritize such research and modes of scientific appraisal and scientific community organization that value it and help to bring it about. This ideal, therefore, seems able to fulfill the political role of the ideal of value-free science. But is it able to fulfill the political role by safeguarding science as a genuine source of knowledge—as the ideal of value-free science aspired to do—or is it able to fulfill the political role by sacrificing science as a genuine source of knowledge? In short, is the ideal of socially responsible science able to fulfill the epistemic role as well as the political role of the ideal of value-free science? This is the question we need to have answered if we are to determine whether the ideal of socially responsible science can fill the position now vacated by the ideal of value-free science. How does the candidate respond? We can begin our interview right here.

The Interview of the Second Candidate

Silence. . . . No answer. . . . Can the candidate be nervous? How else to explain the silence? After all, isn't there an answer to our question already at hand? The idea that politicizing science automatically contaminates it—automatically sacrifices it as a source of genuine knowledge—has already been dealt with. Philosophical naturalists have persuasively argued that social values need not compromise the objectivity of science any more than do other features of scientific communities such as competitiveness, deference to authority, or the desire for credit for one's accomplishments (see, for example, Solomon 2001). What does or does not compromise the objectivity of science, naturalists argue, is an empirical question to be settled by a close examination of scientific practice rather than by an a priori pronouncement regarding the proper conduct of inquiry or the proper composition of scientific communities. Some philosophical naturalists—feminist naturalists—have even argued that social values, as a matter of empirical fact, can be *aids* in the acquisition of objective knowledge—that when these values are allowed to influence science (for example, by motivating particular lines of research or the maintenance of particular social structures) that science can actually be more developed and more empirically adequate than before (see, for example, Antony 1993, 1995; Wylie and Nelson 1998; Campbell 2001; Anderson 1995, 2004). And when we reflect on the effects of feminism in science during the last three decades—the wide-ranging critiques of traditional science in such fields as psychology, sociology, economics, political science, archaeology, anthropology, biology, and medical research, and the new research directions and research results forged in the wake of those critiques—when we reflect on the effects of feminism in science during the last three decades, the arguments of the feminist naturalists seem especially convincing. Egalitarian social values in these cases have seemed to yield better rather than worse science, more objective rather than contaminated science (see Schiebinger 1999 and Creager, Lunbeck, and Schiebinger 2001 for the kinds of wide-ranging changes in science that have occurred due to feminism). Haven't feminist naturalists, then, already provided an answer to our question regarding the epistemic credentials of the ideal of socially responsible science?

But our candidate sees no answer here. Indeed, the above suggestions, interesting and important though they be, are simply too weak to be helpful. After all, the strongest answer to our question that follows from the above is that the ideal of socially responsible science *may* be able to fulfill the epistemic role as well as the political role of the ideal of value-free science. Feminist naturalists certainly cannot now assure us that all the social values recommended by the ideal of socially responsible science will always aid the acquisition of knowledge in all the diverse areas of science. The empirical evidence to support such an assurance is just not there. Even with regard to the so-called "feminist contributions to science" over the last three decades, feminist naturalists would be hard pressed to show that the progress made was in every case the effect of feminist social values rather than other factors. Alison Wylie, for example, has presented survey evidence to show that it was women archaeologists' standpoint as women, not their feminist values (which half the time they denied having), that brought about the dramatic changes in archaeology that began in the late 1980s (Wylie 1997).[5] And Sarah Hrdy only speculates that feminist values were involved in the fundamental rethinking of sexual selection theory that occurred in primatology beginning in the 1970s (Hrdy 1986). And the Biology and Gender Study Group only claims that the eye-opening studies that led to new models of fertilization and sex determination in the 1980s "can be viewed as feminist-influenced critiques of cell and molecular biology": "It should be noted that the views expressed in this essay may or may not be those of the scientists whose work we have reviewed. It is our contention that these research programs are inherently critical of a masculinist assumption with these respective fields. This does not mean that the research was consciously done with this in mind" (The Biology and Gender Study Group 1988, 68, 74n5). And so on.

Of course, if feminist naturalists define "good" social values in science as those that aid the acquisition of objective knowledge, those that are epistemically fruitful—as at least some feminist naturalists seem inclined to do (see Antony 1993 and Campbell 2001)—then the above suggestions are far from weak. For then the ideal of socially responsible science, in allowing only good social values into science, would be allowing only those social values that aid the acquisition of objective knowledge, and would thereby fulfill the epistemic role of the old ideal of value-free science. But this solution has problems of its own. After all, it treats egalitarian social values as merely causally relevant "social factors" or "social biases," on a par with other factors such as competitiveness or the desire for credit or other values such as sexism or racism. All these become possible aids to the acquisition of objective knowledge, and all must be empirically tested to see whether they are. Any of them will do, we are led to infer, if only they can prove their mettle in scientific research.[6] So if, for example, a close comparative study of German medical science before, during, and after the Third Reich discloses that Nazi social values produced the best scientific results, the most abundant and most empirically successful science, then Nazi social values would be good values and should therefore be welcomed into science. Or if such a study discloses that Nazi social values produced a science just as successful as the others, but no better, then it should be a matter of complete indifference whether Nazi social values or the other sciences' values should find their way into science, since none of the values would be justified over the others. And this is remarkable given that one of the main factors that brought about the success of Nazi medical science was the *absence* in it of good social values—for example, the absence (sanctioned by Nazi social values) of moral constraints on human experimentation.

This is not to say that the epistemic success (or failure) of a scientific research project tells us nothing about the justifiability of the social values that guide it. But what it tells us must take into account a great many other factors besides that outcome—for example, which scientists were involved in the project, the level of their talents and training, and the conceptual and material and social resources at their disposal. Factors such as these help to explain the failure of research guided by arguably good social values (such as some of the egalitarian social values guiding Lysenko) and also help to explain the success of research guided by arguably bad social values (such as the racist social values guiding the Nazis). But, of course, moral and legal principles, as well, are relevant to the assessment of the social values that guide scientific research—think of the respect for individual autonomy and self-determination and the Hippocratic Oath's admonition that physicians should "abstain from all intentional wrong-doing and harm" that informed the response to Nazi medical research in the Nuremberg Doctors' Trial and the Nuremberg Code on human experimentation that followed (Katz 1996). And these moral and legal principles, in turn, are themselves informed by factual considerations, including the factual considerations that result from scientific inquiry. What all this shows is that the assessment of the social values that guide a scientific research project, whatever the epistemic outcome of that project, is a complex, multifaceted undertaking. But this still leaves our question concerning the epistemic consequences of these social values in science. What answer can our candidate provide?

Consider, again, Carolyn West's research program. What effects do its egalitarian social values have on it? Do they compromise the objectivity of the knowledge it provides? First, what are these values? They seem to be: "women deserve to live without fear of violence from domestic partners," and "black women deserve the same opportunities as white women to live in such partnerships." These values are well justified both in and out of feminist theory; they should be uncontroversial. Second, what role do these values play in the program? It will be recalled that West makes it a central part of her aim not to perpetuate the stereotype that black Americans are inherently more violent than other ethnic groups. The reason is that this stereotype in people's minds—in the minds of researchers and politicians and service providers, for example—makes it more likely that black women's needs related to domestic violence will be treated less seriously than white women's needs, or even ignored altogether. After all, if violence is perceived as inevitable, as somehow innate or unique to the black culture, intervention efforts are more likely to be perceived as futile. In short, if black women deserve the same opportunities as white women to live in domestic partnerships free of violence—West's egalitarian social value—then the stereotype connecting blacks and violence must not be perpetuated. This means that West's research program as far as empirically possible must highlight the similarities in domestic violence within the black and white communities and seek to explain whatever dissimilarities appear within these communities in terms of social differences such as racism and poverty.[7] But none of this must obscure in any way black women's experiences, since to do so would again be to shortchange black women's needs, and hence fail to provide black women the same opportunities as white women to live in partnerships free of violence. The result, as we have seen, is a dramatic change in research questions; concepts (like the concept of "partner violence" itself); measurement scales and techniques; methods of subject selection; strategies of data collection, analysis, and interpretation; and even methods of publishing and disseminating results.

The upshot: West's research program is controlled through and through by sound egalitarian social values. But it is equally controlled through and through by sound epistemic values. Though the science here is thoroughly politicized, in short, it is not at the expense of its mission to provide genuine knowledge. And this should not be the least bit surprising. After all, research such as West's cannot fulfill its social objectives, cannot effect improvements for battered women in both the black and white communities, unless it does fulfill its epistemic objectives, unless it does get a firm handle on the reality it means to reform. But this means that research such as West's, with its two kinds of interrelated objectives, social and epistemic, shaped by two kinds of values, social and epistemic, should be judged by two kinds of standards, not one—by moral/political standards as well as by epistemic standards. Such research should be found wanting if it fails sound epistemic requirements. But it should also be found wanting if it is shaped by unacceptable social values. How else can science take its rightful place in the forefront of social change?

Where You Take Over the Interview

Has the candidate successfully answered our question?

"Not at all," you exclaim, voice rising. "The answer given is too quick. Indeed, the answer given makes it look as though the epistemic objectives and the social objectives of a research program such as West's can never conflict, so that the social objectives, or the social values that lie behind them, can never contaminate the knowledge produced. But this is far too optimistic. After all, what if the stereotype that black Americans are inherently more violent than other ethnic groups were true?[8] The egalitarian-value-directed research program described would never allow this truth to be discovered, and the ideal of socially responsible science would never allow any less egalitarian research program to be pursued—say, one that straightforwardly investigated the truth of the stereotype by searching for cultural factors associated with violence, cultural factors that differ from one ethnic group to another. So in this case social objectives and epistemic objectives would clearly conflict, and the ideal of socially responsible science would sacrifice the epistemic objectives for the sake of the social. This means that the ideal of socially responsible science cannot be relied on to fulfill the epistemic role as well as the political role of the ideal of value-free science."

"Not so," comes the candidate's reply. (This candidate is not about to concede defeat!) "If the stereotype connecting blacks and violence were true, that truth *could* be discovered with West's program. All the program requires, remember, is that dissimilarities in domestic violence within the black and white communities be explained, *as far as empirically possible,* in terms of social differences such as racism and poverty. The program does not guarantee that any of these explanations will be successful. Indeed, if the stereotype connecting blacks and violence were true, all of these explanations at best would have limited success (depending on whether they also were true), and that would provide (indirect) support for the stereotype. And since a central aim of the program is to make black women's experiences with domestic violence as visible as white women's experiences, the dissimilarities between the two would be made visible as well—just those dissimilarities whose failure to be socially explained would count in favor of the stereotype. So neither the ideal of socially responsible science, nor West's particular research program sanctioned by that ideal, makes knowledge unreachable. Nor do they 'contaminate' the knowledge produced. They simply channel science's search for knowledge in some directions and away from others in response to the needs we present as a society."[9]

"But the 'channeling' runs very deep!" you retort, irritation in your voice. "It affects not only research questions, but also, as we have seen, such aspects of research as concepts, measurement scales and techniques; methods of subject selection; strategies of data collection, analysis, and interpretation; and even methods of publishing and disseminating results. It may even affect other central aspects of the research process such as consideration of the consequences of error and setting acceptable levels of risk (see, for example, Douglas 2000). So the ideal of socially responsible science and the research programs it sanctions may not make knowledge unreachable, nor contaminate the knowledge produced. But they surely slow down the production of knowledge if the channeling is in the wrong direction. If the stereotype connecting blacks and violence were true, for example, the fastest way to discover that truth would doubtless be to investigate the stereotype directly. Not knowing whether the stereotype is true, however, the most plausible way to proceed would be to pursue multiple research programs—the stereotype-focused research program as well as West's egalitarian-value-directed research program, for example. Not only would this be the most efficient way to proceed, but it would also provide valuable comparative assessments of programs in addition to the direct empirical assessments available to each individual program. It would also provide the most thorough assessments, since one program might generate data relevant to another that the other had no access to itself, data with which it nevertheless has to deal. Pursuing multiple research programs would also make more likely the discovery of multiple causal factors and a more complex understanding of the subject at hand. Limiting science to 'socially responsible' research, by contrast, places unnecessary obstacles in the way of science's search for truth."

"That's not true!" gasps the candidate. "Socially responsible research, of course, cannot be guaranteed to produce truth. But neither can socially irresponsible research. Nor can socially responsible research—or socially irresponsible research—be guaranteed to be efficient in its search for truth, or more efficient than the other. We simply cannot say, a priori, what kind of research will produce the best results. If the stereotype connecting blacks and violence were true, for example, would scientists more likely discover that truth, or discover it more quickly or easily, if they explored all plausible ways in which blacks could be inherently disposed to violence, or would they more likely, or more quickly or easily, discover that truth if they explored all plausible social factors that could explain the dissimilarities in violence within the black and white communities? If it be said that the former 'direct' approach would obviously be better, it must be noted that many in the black community would not cooperate with that approach whereas they would cooperate with the latter, socially responsible approach (see West 2002 and 2004 for the 'culture of silence' that has surrounded the problem of domestic violence in the black community, the reasons for it, and the methods that have proven valuable to overcome it). That lack of cooperation would have a profound effect on 'efficiency.' It must also be noted that the latter, socially responsible approach, no less than the other, could make use of multiple research programs, with all the benefits those bestow. So the comparison would not have to be between (as you seemed to suggest) West's program plus the stereotype-focused research program on the one side versus West's program alone on the other. The socially responsible (second) side of the comparison could include, in addition to West's program, any number of other socially responsible alternative or complementary research programs. And, of course, it would matter what all these various research programs were like, which scientists were pursuing them, how much funding they had at their disposal, what background knowledge and conceptual and technological resources they could draw on, etc., etc. The upshot is that you simply cannot assume that limiting science to socially responsible research will slow science down in its search for truth."

"But what if it did?" the candidate continues. "What if the efficiency of research *were* compromised by the restrictions imposed by the ideal of socially responsible science? What grounds are there for saying that these restrictions would then constitute '*unnecessary* obstacles in the way of science's search for truth' when these restrictions—the social values like West's egalitarian values imposed by the ideal of socially responsible science—would be *justified*? Everyone concedes that the value of efficiency in research has its limits, that there are other values, including other social values, that are more important. It might be far more efficient for searching out the truth, for example, if scientists simply ignored the risks to human subjects or society or the environment posed by various lines of research and ethics committees and publishers and funders and the public at large allowed them to do so. But acting in this way would be unconscionable despite the epistemic efficiency it might offer. The ideal of socially responsible science simply extends these constraints already recognized as appropriate for science. In so doing it does not sacrifice science as a genuine source of knowledge, but merely acknowledges that science has other goals and other responsibilities besides its epistemic ones. Thus, it might be more efficient for searching out the truth about domestic violence in the black community if scientists pursued any research they pleased—for example, the stereotype-focused research program in addition to West's approach—irrespective of its effects on the black community. But acting in this way would again be unconscionable. After all, the stereotype-focused research program begins with a characterization of blacks born of prejudice, with no consistent empirical backing, and dignifies it by making it the subject of scientific research. It thereby suggests that the characterization has some plausibility (if it had none, why would scientists bother to investigate it?). And so, the stereotype-focused research program helps to keep the stereotype alive, paradoxically even while it may be accumulating evidence against that stereotype, and as one result (there are others) decreases the likelihood that black women will receive the help they deserve to combat domestic violence. West's program, in contrast, does none of this—is explicitly designed to do just the opposite—even though it also, indirectly, investigates the stereotype. The difference is that West's program aims to help the black community with the knowledge it gathers, and is in an excellent position to do just that. The stereotype-focused research program seems aimed to do just the opposite, and is in an excellent position to do just that. Small wonder that West has received an award from the black community for the work she is doing—the Outstanding Researcher Award from the Institute on Domestic Violence in the African American Community—whereas it is safe to say that the stereotype-focused research program would meet with a very different response."

Where You Come to a Decision

Is it now clear that the ideal of socially responsible science can fulfill the epistemic role as well as the political role of the retired ideal of value-free science?

"If it is," you reply, "that will still not suffice to justify embracing it. The reason the ideal of value-free science retired, remember, was that it could not be put to use. Even what we take to be the greatest science failed to exemplify it, and sociologists and philosophers of science assured us that most science never would exemplify it, never could exemplify it. In short, the ideal of value-free science failed to be a viable ideal, useful for actual science. Is the ideal of socially responsible science similarly inapplicable?"

"Not at all," boasts the candidate triumphantly. "Unlike value-free science, socially responsible science is possible. Indeed, it exists. As noted at the outset, feminist scientists such as West are among the scientists who are doing it. This does not mean that all that we currently consider the greatest science is socially responsible science. That has to be determined on a case by case basis, and feminist scientists and historians and philosophers of science, among others, have already done some of this work. But it does mean that there actually are concrete models now available to other scientists that help to show them what the ideal of socially responsible science amounts to and how it can be put into practice. And it also means that . . ."

Here the candidate is interrupted by you, waxing impatient: "Questions concerning values—including the 'values that meet the needs of society,' the values that the ideal of socially responsible science aims to entrench in science—are highly controversial. Even feminists, who agree on so many things, are far from agreement concerning what their egalitarian social values amount to and how they can best be put into practice—for example, exactly what a gender equal society would be like and how it should be pursued. So isn't the ideal of socially responsible science just as inapplicable as the ideal of value-free science, since no one can agree concerning what would satisfy it—concerning which values would meet the needs of society?"

"You are not *listening*," replies the candidate. "Of course there is disagreement concerning values, including the 'values that meet the needs of society.' But there is also crucially important agreement, especially concerning the concrete issues that affect people's day-to-day lives. Regarding West's research, for example, it is uncontroversial that women deserve to live without fear of violence from domestic partners, the value that underlies West's research. But it is equally uncontroversial that women deserve to live without fear of rape, sexual harassment, incest, and other forms of violence directed at women, and that women deserve equal educational opportunities with men, equal employment opportunities with men, equal opportunities for health care, and so on. These values are not only uncontroversial in Western cultures, they are attested to in the policy declarations and activities of such international organizations as the United Nations, the International Labour Organization, the World Health Organization, and Amnesty International. These are the kinds of shared values that motivate and inform feminist research in such fields as psychology, sociology, economics, political science, archaeology, anthropology, biology, and medical research. And this is the kind of research that exemplifies the ideal of socially responsible science."

"So your question should be," the candidate continues, "not *is* the ideal of socially responsible science applicable to real science under real, that is, our current, social conditions, but *how extensively* is this ideal applicable to this science under these conditions. That is to say, can the shared social values that shape the research of feminist scientists come to shape the research of other scientists as well, and can other social values that meet the needs of society but do not now shape research be added to them? These are large-scale empirical questions, but fortunately there is enough empirical evidence currently available to at least begin to answer them. Certainly the long-term flourishing of feminism in some fields—for example, primatology, cultural anthropology, paleontology, and developmental biology—and its recent growth in others—for example, archaeology—give cause for optimism. Primatology is a particularly good example (see, for what follows, Fedigan 2001). This field has wholeheartedly embraced feminist ways of doing research—in pursuing research that rescues female primates from their previous second-class status in the theoretical understandings of the field (as merely mothers, as merely passive resources for males), in pursuing research that answers questions of importance to women (regarding male parenting roles or the evolution of female sexuality, for example), in pursuing research that uses new female-friendly conceptual tools (for example, sampling methods that more readily include females), and so on. Primatology has embraced such feminist ways of doing research even though very few of its practitioners see themselves as feminists, and even though the standard attitude of these practitioners is that politics does not belong in science. More significantly, the reasons these practitioners give for doing so—that it makes for better *science*, that it is *scientifically* right to consider questions from a female as well as a male perspective, to research issues of concern to women as well as men, and about females as well as males—give cause to be hopeful that further applications of the ideal of socially responsible science are possible."

"But other fields tell a different story!" you interject.

The candidate falls silent for a moment. Then: "Other fields have made some of the same changes as primatology, but only under duress. U.S. medical research, for example. Only since 1993, when Congress passed the National Institutes of Health Revitalization Act that mandated the inclusion of women and minority men in publicly funded U.S. biomedical research and made funding contingent on that inclusion, has the neglect of females in both basic and clinical research been curtailed. Earlier initiatives—for example, NIH's 1986 guidelines requiring grant applications to include female subjects in medical testing and research—were generally ignored (Rosser 1994; Schiebinger 1999). And still other fields have made few if any changes—economics, for example, in which women's needs and priorities in the family as well as the larger society remain invisible or inadequately treated (Nelson 1996a, 1996b; Waring 1997). Do these cases show that the ideal of socially responsible science is of limited applicability? Not at all. The case of U.S. medical research shows that economic incentives—not only public funding but very possibly also tax incentives for industry-funded science and conditions on the tax exempt status of foundation-funded science—can be a powerful method to bring about socially responsible science. The case of economics shows that the need in some areas has never been greater. And both cases show that much hard work will have to be done to determine for each of the various fields of science how best to achieve what the ideal of socially responsible science recommends. But this is the kind of work that makes sense only if, and after, the ideal of socially responsible science is adopted. For this work will answer the question, not how extensively is the ideal of socially responsible science applicable, but how can it be made more applicable."

Do you have any more questions? Or is the interview over? And if so, what is its outcome? Has the ideal of socially responsible science shown both that it can fulfill the epistemic and political roles of the old ideal of value-free science and that it will actually get the job done, not fail to apply itself? Has it shown, in short, that it merits the position now vacated by the ideal of value-free science? Since I have put forward this candidate (and also argued in its defense) you should already know where I stand. But you have followed the interview as closely as I and you have posed most of the questions, so you play a role here, too. The decision, I think, now depends on you.

NOTES

This paper was completed while I was a visiting fellow at the University of Pittsburgh's Center for Philosophy of Science. I gratefully acknowledge the research support of both the Center and the University of Notre Dame.

1. This was not the only way, as I shall make clear later.

2. Certainly, the sorts of questions Longino says must be answered in order to complete her analysis—for example, "In determining what counts as inappropriate exclusion of dissenting perspectives, does it matter what kind of issue is involved?" and "What bearing should greater cognitive authority have on the attribution of intellectual authority, understood as the capacity to participate in critical discussion and thus to contribute to critical understanding?" (Longino 2002, 133)—certainly these sorts of questions do not seem answerable by yet another round of reflection on the meaning of "knowledge" and related terms.

3. This may be a reason to deny that what PETERS produces is knowledge after all, that is, to deny that Longino's social value management ideal of science fulfills the epistemic role of the ideal of value-free science.

4. Of course Longino might say at this point that if the requisite political movements or analytic methods or mathematical resources or instrumental technologies or funding or staffing or family supports were not there to aid the underprivileged ones in PETERS, then "all relevant perspectives" were not there in PETERS either. Hence, the conclusion to be drawn is not that the social value management ideal of science fails to fulfill the political role of the ideal of value-free science, but that the social value management ideal of science has not been provided with a genuine test case to see if it does. But such a response

on Longino's part would threaten to make her candidate's fulfillment of the political role of the ideal of value-free science true by definition (of "all relevant perspectives are represented"). It would also threaten to make her candidate unrealizable in practice—just the problem that caused the retirement of the ideal of value-free science.

5. Note, however, that Wylie qualifies this stand in Wylie and Nelson (1998). There she speaks of the changes in archaeology as having been brought about by "a standpoint of gender sensitivity—a grass-roots feminist sensibility" that "reflects the indirect influence of the second wave women's movement" (1998, 7).

6. Antony (1993) and Campbell (2001) classify together sexism and racism along with feminist social values as "biases." They then make a distinction between "good biases"—those that "facilitate the gathering of knowledge," that is, those that "lead us to the truth"—and "bad biases"—those that "lead us *away* from the truth." In short, "*we must treat the goodness or badness of particular biases as an empirical question*" (Antony 1993, 215, emphasis hers; and cf. Campbell 2001, 196, who quotes Antony approvingly, although he tries for a more elaborate kind of naturalism in 1998). Solomon (2001) also classifies together sexism, racism, and egalitarian social values, now as "ideology," and goes on to group together these ideological social factors ("decision vectors" or causes for theory choice) with other "nonempirical" decision vectors such as birth order, desire for credit, deference to authority, and competitiveness. But for her, an equal distribution of such nonempirical decision vectors among competing theories is what generally helps to produce "normatively appropriate" science.

7. One example West presents: "Black feminist thought can make a significant contribution by keeping the focus on historical perspectives. During slavery and well into reconstruction, Black women witnessed their husbands, fathers, sons, and brothers being abducted by slave owners, police officers, and Klansmen. For the contemporary Black woman, having her partner arrested may be reminiscent of these earlier historical traumas. Although she wants the violence to stop, she may be reluctant to thrust her batterer into a system that is discriminatory, hostile, and overcrowded with Black males. Batterers realize this and will often use this history to further manipulate their partners. Black feminists recommend that this history be acknowledged while simultaneously holding African-American men accountable for their abuse" (2002, 229).

8. Note that West (2002) begins by explaining that the studies conducted to date present a "contradictory" picture of racial differences in domestic violence. "In summary, some researchers found similar rates of partner violence across racial groups. . . . In contrast, other investigators discovered that Black women, when compared to their White counterparts, were significantly more likely to sustain and inflict aggression. Moreover, they were more likely to be victims of severe violence. This pattern was reported at every stage on the relationship continuum" (218). West's research program is a socially and epistemically sophisticated way to deal with this contradictory situation. What is now being suggested is that there may be other ways, and perhaps even better ways, to deal with it that should also be considered.

9. We can think of West's program as a Lakatosian research program (see Lakatos 1970). West's denial of the stereotype that black Americans are inherently more violent than other ethnic groups is part of the "hard core" of the program. Her instructions to highlight similarities and to explain away dissimilarities between the black and white com-

munities are part of the "negative heuristic" of the program that protects the hard core from refutation. And her instructions regarding how to do this—e.g., to revise concepts such as "'partner violence'" to uncover similarities and to formulate hypotheses to explain dissimilarities in terms of social factors such as racism and poverty—are part of the "positive heuristic" of the program. Finally, although Lakatos never considered social values as playing a legitimate role within scientific research programs, what motivates West's program are her egalitarian social values. What the candidate has just been saying is that there are conditions under which it will be rational to abandon (to consider "refuted") West's research program, conditions that Lakatos tried to describe in detail. Notice, however, that the abandonment of West's research program would not necessarily justify the abandonment ("refutation") of West's egalitarian social values, for reasons that were made clear in the preceding section. For example, if it were concluded that the stereotype about black Americans is true (which is the denial of the program's hard core), say because violence is inherent in black culture, it would not follow that black women do not deserve the same opportunities as white women to live without fear of violence from domestic partners. See for example the complicated debate about the relationship between feminism and multiculturalism in Okin 1999.

REFERENCES

Anderson, Elizabeth. 1995. "Knowledge, Human Interests, and Objectivity in Feminist Epistemology." *Philosophical Topics* 23 (2): 27–58.

———. 2004. "Uses of Value Judgments in Science: A General Argument, with Lessons from a Case Study of Feminist Research on Divorce." *Hypatia* 19 (1): 1–24.

Antony, Louise. 1993. "Quine as Feminist: The Radical Import of Naturalized Epistemology." In *A Mind of One's Own: Feminist Essays on Reason and Objectivity*, ed. Louise Antony and Charlotte Witt, 110–53. Boulder: Westview.

———. 1995. "Sisters, Please, I'd Rather Do It Myself: A Defense of Individualism in Feminist Epistemology." *Philosophical Topics* 23 (2): 59–94.

Bernal, John Desmond. 1971. *Science in History*, vol. 2: *The Scientific and Industrial Revolutions*. Cambridge, MA: MIT Press.

Biology and Gender Study Group. 1988. "The Importance of Feminist Critique for Contemporary Cell Biology." *Hypatia* 3:61–76.

Bleier, Ruth. 1984. *Sex and Gender*. New York: Pergamon Press.

Campbell, Richmond. 1998. *Illusions of Paradox: A Feminist Epistemology Naturalized*. Lanham, MD: Rowman and Littlefield.

———. 2001. "The Bias Paradox in Feminist Epistemology." In *Engendering Rationalities*, ed. Nancy Tuana and Sandra Morgen, 195–217. Albany: State University of New York Press.

Creager, Angela N., Elizabeth Lunbeck, and Londa Schiebinger, eds. 2001. *Feminism in Twentieth-Century Science, Technology, and Medicine*. Chicago: University of Chicago Press.

Di Leonardo, Micaela, ed. 1991. *Gender at the Crossroads of Knowledge: Feminist Anthropology in the Postmodern Era*. Berkeley: University of California Press.

Douglas, Heather. 2000. "Inductive Risk and Values in Science." *Philosophy of Science* 67 (4): 559–79.

Dupré, John. 2007. "Fact and Value." In *Science and Values*, ed. John Dupré, Harold Kincaid, and Alison Wylie. Oxford: Oxford University Press.

Eichler, Margrit. 1980. *The Double Standard: A Feminist Critique of Feminist Social Science*. New York: St. Martin's Press.

———. 1988. *Nonsexist Research Methods: A Practical Guide*. Boston: Allen and Unwin.

Elkana, Yehuda. 1982. "The Myth of Simplicity." In *Albert Einstein: Historical and Cultural Perspectives*, ed. Gerald Holton and Yehuda Elkana, 205–52. Princeton: Princeton University Press.

Fausto-Sterling, Anne. 1992. *Myths of Gender*. 2nd ed. New York: Basic Books.

Fedigan, Linda Marie. 2001. "The Paradox of Feminist Primatology: The Goddess's Discipline?" In Creager, Lunbeck, and Schiebinger 2001, 46–72.

Gilman, Sander L. 1993. *Freud, Race, and Gender*. Princeton: Princeton University Press.

Haraway, Donna. 1989. *Primate Visions: Gender, Race, and Nature in the World of Modern Science*. New York: Routledge.

Harding, Sandra, ed. 1993. *The "Racial" Economy of Science: Toward a Democratic Future*. Bloomington: Indiana University Press.

———. 1998. *Is Science Multicultural?: Postcolonialisms, Feminisms, and Epistemologies*. Bloomington: Indiana University Press.

Hrdy, Sara Blaffer. 1986. "Empathy, Polyandry, and the Myth of the Coy Female." In *Feminist Approaches to Science*, ed. Ruth Bleier, 119–46. New York: Pergamon Press.

Hubbard, Ruth. 1990. *The Politics of Women's Biology*. New Brunswick, NJ: Rutgers University Press.

Katz, Jay. 1996. "The Nuremberg Code and the Nuremberg Trial: A Reappraisal." *Journal of the American Medical Association* 276 (20): 1662–66.

Keller, Evelyn Fox. 1985. *Reflections on Gender and Science*. New Haven: Yale University Press.

———. 1992. *Secrets of Life, Secrets of Death: Essays on Language, Gender, and Science*. New York: Routledge.

Kitcher, Philip. 2001. *Science, Truth, and Democracy*. New York: Oxford University Press.

Knorr-Cetina, Karin. 1981. *The Manufacture of Knowledge*. Oxford: Pergamon Press.

Knorr-Cetina, Karin, and Michael Mulkay, eds. 1983. *Science Observed: Perspectives on the Social Study of Science*. London: Sage.

Kramarae, Cheris, and Dale Spender, eds. 1992. *The Knowledge Explosion*. New York: Teachers College Press.

Lakatos, Imre. 1970. "Falsification and the Methodology of Scientific Research Programmes." In *Criticism and the Growth of Knowledge*, ed. Imre Lakatos and Alan Musgrave, 91–196. Cambridge: Cambridge University Press.

Latour, Bruno. 1987. *Science in Action: How to Follow Scientists and Engineers through Society*. Cambridge, MA: Harvard University Press.

Longino, Helen. 1987. "Can There Be a Feminist Science?" *Hypatia* 2 (3): 51–64.

———. 1990. *Science as Social Knowledge: Values and Objectivity in Scientific Inquiry*. Princeton: Princeton University Press.

———. 2002. *The Fate of Knowledge*. Princeton and Oxford: Princeton University Press.

Merchant, Carolyn. 1980. *The Death of Nature: Women, Ecology, and the Scientific Revolution*. San Francisco: Harper and Row.

Nelson, Julie A. 1996a. *Feminism, Objectivity and Economics*. London: Routledge.

———. 1996b. "The Masculine Mindset of Economic Analysis." *Chronicle of Higher Education* 42 (42): B3.

Okin, Susan. 1999. *Is Multiculturalism Bad for Women?* Princeton: Princeton University Press.

Potter, Elizabeth. 2001. *Gender and Boyle's Law of Gases*. Bloomington: Indiana University Press.

Proctor, Robert N. 1991. *Value-Free Science?: Purity and Power in Modern Knowledge*. Cambridge, MA: Harvard University Press.

Putnam, Hilary. 2002. *The Collapse of the Fact/Value Dichotomy*. Cambridge, MA: Harvard University Press.

Rosser, Sue. 1986. *Teaching Science and Health from a Feminist Perspective: A Practical Guide*. New York: Pergamon Press.

———. 1990. *Female-Friendly Science: Applying Women's Studies Methods and Theories to Attract Students*. New York: Pergamon Press.

———. 1994. *Women's Health—Missing from U.S. Medicine*. Bloomington: Indiana University Press.

———, ed. 1995. *Teaching the Majority: Breaking the Gender Barrier in Science, Mathematics, and Engineering*. New York: Teachers College Press.

———. 1997. *Re-engineering Female-Friendly Science*. New York: Teachers College Press.

Ruse, Michael. 1999. *Mystery of Mysteries: Is Evolution a Social Construction?* Cambridge, MA: Harvard University Press.

Schiebinger, Londa. 1999. *Has Feminism Changed Science?* Cambridge, MA: Harvard University Press.

Shapin, Steven, and Simon Schaffer. 1985. *Leviathan and the Air-Pump: Hobbes, Boyle, and the Experimental Life*. Princeton: Princeton University Press.

Solomon, Miriam. 2001. *Social Empiricism*. Cambridge, MA: MIT Press.

Spanier, Bonnie B. 1995. *Im/partial Science: Gender Ideology in Molecular Biology*. Bloomington: Indiana University Press.

Waring, Marilyn. 1997. *Three Masquerades: Essays on Equity, Work, and Hu(man) Rights*. Toronto: University of Toronto Press.

West, Carolyn. 2002. "Black Battered Women: New Directions for Research and Black Feminist Theory." In *Charting a New Course for Feminist Psychology*, ed. Lynn Collins, Michelle Dunlap, and Joan Chrisler, 216–37. Westport, CT: Praeger.

———, ed. 2004. *Violence in the Lives of Black Women: Battered, Black, and Blue*. New York: Haworth Press.

Wylie, Alison. 1997. "The Engendering of Archaeology: Refiguring Feminist Science Studies." In *Women, Gender, and Science: New Directions*, ed. Sally Kohlstedt and Helen Longino. *Osiris* 12:80–99.

Wylie, Alison, and Lynn Hankinson Nelson. 1998. "Coming to Terms with the Value(s) of Science: Insights from Feminist Science Scholarship." Paper delivered at the Workshop on Science and Values, Center for Philosophy of Science, University of Pittsburgh.

5

SCIENTIFIC VALUES AND
THE VALUES OF SCIENCE

JAY F. ROSENBERG
University of North Carolina

W HEN IT COMES TO SAYING SOMETHING USEFUL about science and
values, one of the first things we need to do is to locate science. Richard
Rorty, for example, distinguishes Platonists, who hold that science fundamentally consists of procedures for getting our representational "schemes" into
increasingly better touch with the world's "contents," from Baconians, who
"call a cultural achievement 'science' only if they can trace some technological advance, some increase in our ability to predict and control, back to that
development" (Rorty 1991, 5, 47).[1] Roughly, Platonists view scientific inquiry as
a source of epistemic authority, licensing ontological claims such as that "there
are no [astrological] planetary influences out there, whereas there really *are*
atoms out there" (Rorty 1991, 5), while Baconians think of scientific inquiry in
terms of empowerment, as better situating us, in Bacon's own words, "to relieve
and benefit the condition of man." In more traditional terms, we might say that
Platonists put theoretical science at center stage while Baconians give pride of
place to applied science. In any case, when Platonists talk about science, they
are characteristically thinking of a determinate cognitive discipline. Baconians,
in contrast, are characteristically thinking of a diverse family of social practices.

In first approximation, both Janet Kourany and Helen Longino are Baconians; I am a particular sort of heretical Platonist.

Rorty sees these two conceptions of science as competing and mutually exclusive. I've suggested elsewhere (Rosenberg 1993) that they are compatible and complementary, and I shall shortly have more to say on the matter. But the first essential point is that values are implicated on both conceptions, although in different ways. Kourany in effect recognizes this when she observes that the ideal of value-free science played both an epistemic role and a political role, suggesting a way to achieve both objective knowledge and social reform. She traces some of the history of this ideal, but we can already see that, even on the most austere epistemic reading of "science," "value-free" turns out to be a misnomer, for a commitment to intellectual and methodological principles can reflect values no less than do commitments to moral and political principles. Thus Longino is not at all reluctant to speak of cognitive, epistemic, or even theoretical values, although she tends to privilege talk in terms of "virtues" or "heuristics."

What I take to be at issue between Baconians like Longino and Kourany and Platonists like me, then, is not whether values are both respected in and subserved by scientific research, but rather the proper way to understand the relationship between the cognitive-epistemic interests characteristically reflected in the values emphasized by Platonists and the sociopolitical interests characteristically reflected in those emphasized by Baconians. As I read them, Kourany and Longino see the two groups of values as mutually influencing and interdependent. In contrast, I see them in an order of priority. Baconians observe that diverse interests can be and are promoted both in and by applying the methods and results of scientific inquiry to real world (practical) problems. That sounds like a platitude, and in any event, it is not something that I would want to deny. But we Platonists have always insisted that inquiry counts as scientific in the first place only if it exemplifies a determinate set of intellectual virtues and is responsible to determinate methodological constraints. Thus a Platonist, but typically not a Baconian, would be prepared to frame the distinction by contrasting constitutive values, intrinsic to scientific practices and theorizing per se, with collateral values, concerning the extrinsic instrumental implications of those practices and their outcomes.

I like to compare science with soccer. Goals are the one constitutive value of soccer. The point of soccer is scoring goals. Insofar as you are playing soccer, the reasons for what you do all have to do with scoring goals: You are trying to help your team to score goals, and you are trying to help prevent the opposing team from scoring goals. In contrast, there are many unrelated reasons for joining a soccer team and for playing soccer. You might be interested in improving your physical fitness and stamina. You might be seeking companionship or a pleasant recreational diversion or trying to measure up to your parents' expectations or hoping to impress your friends with your sports prowess. If you are sufficiently skilled, you might even be aiming at a career as a professional soccer player and then perhaps also at wealth and fame. Analogously, there are many unrelated reasons for forming or sponsoring soccer teams or leagues and for supporting or promoting soccer-playing activities. Businesses might be seeking higher profits through effective advertising and improved public goodwill. Cities or states or nations might be interested in increasing tourism and stimulating the economy, or they might be aiming at enhanced civic, regional, or national pride or patriotism, or trying to distract a potentially rebellious public from political scandals and corruption. All of these, and indefinitely many more, can be collateral values of soccer—but none of them could be collateral values of soccer if there wasn't independently such a thing as the game of soccer. At least in the case of soccer, constitutive values are prior to collateral values in the order of understanding.

On my view, this is equally true in the case of science. Particular cognitive-epistemic values are constitutive for scientific inquiry per se, and consequently they are prior to whatever sociopolitical values are, collaterally, subserved by such inquiry. It is a mistake to suppose that one can identify something properly called "scientific inquiry" and then inquire as to its goal. Inquiry is scientific in part by virtue of its participants having a particular goal, just as kicking a ball around becomes playing soccer in part by virtue of its participants having a particular goal. Inquiry in general, and scientific inquiry in particular, are goal-constituted activities.

As I understand her, Kourany would deny this. And if this is right, there will be a dimension of free play regarding what sort of "science" we want to have. Thus Kourany thinks that it makes sense to advocate an ideal of socially responsible science that "directs scientists to include only specific social values in science, namely the ones that meet the needs of society" (see chapter 5, this volume). One wonders if she would be prepared to go as far as Philipp Frank once did in claiming that since the observed facts, simplicity, and common sense do not uniquely determine a choice of theories, "we can still require their fitness to support desirable moral and political doctrines" (see Frank 1956). Or perhaps it is not the choice of theories that is at issue for Kourany.

Longino's views are more complicated. In *Science as Social Knowledge*, she indeed explicitly sets the question, "What are the goals of scientific activity, realization of which determine the success of that activity and criteria by which to measure success?" (Longino 1990, 18). But rather than proceeding to identify some such "internal" goals, as she puts it, as constitutive for science, in my sense of the term, she proceeds to argue that we should remain "noncommittal about the specific internal goods of scientific inquiry," inter alia, because "modern philosophical work on scientific methodology has, in fact, been guided by quite different conceptions of what a good theory is or does" (19).

This is not the occasion for a detailed exploration of Longino's subsequent discussions of "logical positivism" and "wholism" (*sic*). It suffices to note that, as she sees it, two major conceptions of the goals of scientific practice thereby become salient: "the construction of comprehensive accounts of the natural world," and "the discovery of truth about the natural world" (Longino 1990, 32, 33). As she parses it, the first of these conceptions is evidently Baconian; scientific inquiry is "the search for descriptions of the natural world that allow for the prediction and control of an increasing number of its aspects" (Longino 1990, 32). The second is Platonistic; scientific inquiry has "the goal of faithfully representing nature and not that of expanding explanatory paradigms or our technological dominion over nature" (Longino 1990, 34). And Longino not only sees these as conflicting conceptions as, but, more fundamentally, as reflecting "a tension within science itself between its knowledge-extending mission ('explanatory growth') and its critical mission ('testing, retesting, rejecting, and reformulating hypotheses')" (34).

Longino herself adopts a relaxed nonhierarchical pluralism with respect to "theoretical virtues or values," that is, the qualities or properties of a theory, model, or hypothesis that qualify it as praiseworthy or plausible or worthy of acceptance. Such feminist-inspired virtues as novelty, ontological heterogeneity, complexity or mutuality of interaction, applicability to human needs, and empowerment, she holds, are "epistemologically on a par" with the traditional theoretical virtues of conservatism, simplicity, explanatory power, generality, fruitfulness, and refutability. Both sets of values have heuristic, but not probative, power; both can be politically valenced in appropriate contexts.

But this puts at least one important cart before its corresponding horse. Longino writes that science itself has various missions, but what is this "science itself"? Wasn't that just what we were in the process of trying to locate? My suspicion is that we are tacitly being invited to adopt what one might call a commendatory reading of "science" and "scientific." On this use, "scientific research" is essentially equivalent to "epistemically responsible research," and calling a theory or hypothesis "scientific" just amounts to asserting its intellectual respectability. This is an understandable development from the historical role played by science as a counterforce to dogma and superstition during the Enlightenment, in consequence of which being scientific itself assumed the status of an intellectual virtue or value. But it also sets the stage for something rather like a fallacy of illicit conversion, confusing the view that scientific inquiry is intellectually virtuous with the thesis that all intellectually virtuous inquiry is scientific.

My own heretical Platonism is neither logical positivism nor holism (although it shares significant convictions with both views), but rather a particular form of scientific realism. Longino recognizes this as a possible third position and characterizes it "most crudely" as "the claim that the theories of the 'mature' sciences (for example, physics) are approximately true and that the more recent theories of such sciences approach truth more closely than earlier theories" (Longino 1990, 29). In contrast, as I understand scientific realism, a scientific realist is someone who accepts the following four claims:

1. The controlling aim of scientific inquiry is to provide satisfactory explanatory accounts of phenomena.

2. Such explanatory accounts sometimes posit the existence of theoretical entities.

3. The fact that an account which posits the existence of theoretical entities satisfactorily explains particular phenomena is a good reason for believing that the posited entities exist.

4. All such explanatory accounts are always empirically defeasible.

As I understand scientific realism, however, a scientific realist need not accept either of the following claims:

a. "The terms of a mature scientific theory typically refer." (Richard Boyd)

b. "The theories accepted in a mature science are typically approximately true." (Hilary Putnam)

That is an important part of what makes my Platonism heretical.

For reasons that I shall explain shortly, I would resist any characterization of scientific realism that made use of unexplicated notions of reference or truth. My own characterization is, I think, consistent with Wilfrid Sellars's claim that "To have good reason to accept a theory is to have good reason to believe that the entities it postulates are real"; but since I think that "real" is ultimately in the same boat as "reference" and "truth," I wouldn't want to put it that way, either. Sellars's *scientia mensura* is better: "In the dimension of describing and explaining the world, science is the measure of all things, of what is that it is, and of what is not that it is not" (Sellars 1963, §42). This may first remind you of Protagoras, but I hope that it also reminds you of Aristotle: Truth is "to say of what is that it is, or of what is not that it is not."[2]

On my view, then, the constitutive goal of scientific inquiry is the explanatory accommodation of experience. This is a complicated business, since experience is a complicated business. On the one hand, we find ourselves with experiences. There is a nonprocedural immediacy attaching to perceptual experiences that is one ultimate source of the claim to objectivity carried by our concepts and theories. (The other is the person-neutral intersubjectivity correlative to the requirement that observations and measurements be repeatable.) On the other hand, our concepts and theories in part determine with what experiences we can find ourselves, and our concepts and theories are malleable. We can have good reasons to deliberately modify or abandon them.

The notion of "explanatory accommodation" adverts to the fact that inquiry does not, because it cannot, occur in an epistemic vacuum. Inquiry necessarily occurs within a context of what, for the purpose of conducting it, is provisionally assumed to be known, and it always comes with an agenda, that is, there is always something that a particular course of inquiry aims at finding out, a determinative question—paradigmatically, a "why" question—to be answered. A course of inquiry ends when we are satisfied that we can answer its determinative question. What we count as a satisfactory answer varies from question to question. That's arguably the only theory of explanation that we have or need, although one can thematize significant special cases, for example, perceptual observation and theory succession in physics.[3]

The process I have just sketched is in fact what Peirce called "abduction," but it is a mistake to describe it as a form of inference, for example, "inference to the best explanation," a characterization that Longino uncritically adopts. "The consequence of this form of scientific realism," she concludes, "is a broadening of the category of *evidence* to include not just empirical data but the explanatory virtues of the theory with respect to that data" (Longino 1990, 30). This strikes me as misleading at best. Evidence is one thing; reasoning from and about evidence is another. Be that as it may, Longino finds the strategy of appealing to explanatory considerations problematic on quite general grounds: "Put most simply it is that explanatory arguments at best demonstrate the plausibility of their conclusions" (1990, 30).

If all that this meant was that any explanatory account always remains defeasible—my point 4 above—then, of course, I would have no quarrel with it. But Longino appears to have something more substantial in mind. Her complaint in essence is that explanatory considerations are ill-suited to serve as criteria of truth. "Surely h_1 is a better explanation of e than h_2, other things being equal, if h_1 is true and h_2 is not. If truth is among the criteria we use to determine explanatory superiority, then explanatory superiority cannot in turn be used as an independent criterion of truth" (Longino 1990, 30). Now, as it happens, my sort of heretical Platonist can straightforwardly reply that truth is *not* among the criteria that we use, or even could use, to determine explanatory superiority, but that would plainly be only the opening move of a more extensive dialectic. For my purposes, the most useful way to proceed at this point is, in fact, to embark upon just that dialectic, but unfortunately Longino herself has little more to say about the scientific realist conception of science, instead citing and deferring to the critical initiatives of others. In the balance of this essay, I shall consequently articulate and defend my particular version of scientific realism, not through further dialogue with Longino, but rather by engaging what is arguably the most significant and challenging among those critical initiatives, namely, Bas van Fraassen's "constructive empiricism."

To begin, we need to take a quick look at some of the distinctions that a constructive empiricist wants to stress. One central distinction is between believing a theory and accepting a theory. It is absolutely crucial to constructive empiricism that acceptance is something other than belief. How does the distinction go?

> To believe a theory is to believe that one of its models correctly represents the world. A theory may have isomorphic models; that redundancy is easily removed. If it has been removed, then to believe the theory is to believe that exactly one of its models correctly represents the world. Therefore, if we believe of a family of theories that all are empirically adequate, but each goes beyond the phenomena, then we are still free to believe that each is false and, hence, their common part

> is false. For that common part is phrasable as: one of the models of one of those theories correctly represents the world. (van Fraassen 1984, 252)

To believe a theory, in short, is to believe that it is true—all of it, including its entirely theoretical part—where being true, in turn, is a matter of "correctly representing the world." "Acceptance of a theory," however, "may properly involve something less (or other) than belief that it is true" (van Fraassen 1984, 250). Van Fraassen's elucidation thus invokes a second distinction, namely, between a theory's correctly representing the world and its being "empirically adequate."

The scientific theories that we're interested in here tell stories about the world that go beyond the limits of observation. What is observable, van Fraassen argues, is necessarily limited in various ways. "The most general limit is that experience discloses to us no more than what has *actually* happened to us so far" (van Fraassen 1985, 253). Other limits derive from the de facto constitution of members of the scientific community, for example, our specific sensory capacities. What they are in detail is itself an empirical matter. A theory is empirically adequate if it correctly represents the observable phenomena. More technically:

> · A theory provides, among other things, a specification (more or less complete) of the parts of its models that are to be direct images of the structures described in measurement reports. . . . [Let] us call them "empirical substructures." The structures described in measurement reports we may . . . call "appearances." A theory is *empirically adequate* exactly if all appearances are isomorphic to empirical substructures. (van Fraassen 1984, 257)

From the epistemic perspective, then, we can treat a theory in two different ways: we can accept it as empirically adequate or believe it to be true (van Fraassen 1984, 258). Constructive empiricism adopts the first of these attitudes. A theory is acceptable if it "saves the phenomena," and "the phenomena are saved when they are exhibited as fragments of a larger unity" (van Fraassen 1984, 256). This larger, theoretical unity might also have prima facie consequences that go beyond the observable phenomena, but "our practical commitments require no credence in the more theoretical parts of the theory, and our commitments to research directions or to a direction for further theoretical exploration also need not involve that" (van Fraassen 1985, 295).

My most fundamental disagreement with van Fraassen derives from my inability to make sense of such notions as "how the world is," and so of "the *truth* about the way the world is," independently of an account of the epistemology of scientific inquiry. But before I explain why I reject constructive empiricism, it will be useful to look at why van Fraassen accepts it. His basic reason seems to be that, when the chips are down, only constructive empiricism conscientiously remains true to the essential constitutive principle of empiricism as such. One can be a scientific realist, he suggests, only if one is willing to forgo empiricism, strictly and properly construed.

> [The] core doctrine of empiricism is that experience is the sole source of information about the world and that its limits are very strict. For an example of such a limit we may take our temporal finitude: it is not possible to have a guarantee about the future on the basis of our experience so far—otherwise we could indeed have opinions today which are not liable to modification in the course of future experience. I explicate the general limits as follows: *experience can give us information only about what is both observable and actual*. We may be rational in our opinions about other matters . . . but any defense of such opinions must be by appeal to information about matters falling within the limits of the deliverances of experience. (van Fraassen 1985, 253)[4]

Wilfrid Sellars once characterized "empiricism" as an "accordion word"— one whose sense expands and contracts and, in the process, makes a great deal of philosophical music. Van Fraassen's explication of empiricism belongs among the austere versions. Mine is more expansive. I would certainly not quarrel with the claim that experience is the sole source of information about the world—but "experience" and "information" are also accordion words, and my take on them tends to be considerably more accommodating than van Fraassen's. Crucially, however, we will also need to say something about "the world."

Van Fraassen knows very well, of course, that the notion of observability can coherently be construed in ways that allow for it to be literally correct to say that we observe, for instance, electrons, neutrinos, and black holes. To the best of my knowledge, no one has yet claimed to have observed a free quark, but the project of trying to do so is evidently one that makes sense to the relevant practitioners. In the present debate, however, the operative notion of observability is a function of certain empirically determinable limitations of human beings qua biological organisms.[5] It is in this restricted sense of "observability" that scientific theories tell stories that go beyond the limits of observability. The crux of the debate, according to van Fraassen precisely concerns the role of such limits: "Why should its limits be given any epistemological role at all?

Why should the range of accessible possible evidence play a role, as well as what the actual evidence is?" (van Fraassen 1985, 254).

His answer is, in essence, that he cannot imagine a reasonable alternative.

> If we choose an epistemic policy to govern under what conditions, and how far, we will go beyond the evidence in our beliefs, I could not envisage a nonextreme rational policy that would make these boundaries independent of our opinions about the range of possible additional evidence. . . .
>
> Suppose that nothing except evidence can give justification for belief. However flexibly this is construed, it means that we can have evidence for the truth of a theory only via evidential support for its empirical adequacy. The evidence then still provides some reason for believing in the truth . . . but the additional belief is supererogatory. However, not everyone shares my view that what I called the pragmatic virtues provide no independent reason for belief. (van Fraassen 1985, 254, 255)

Although the terminology is contentious, I have already indicated that I find myself in the latter company. ("Evidence" and "justification" are two more of those accordion words.) That makes my empiricism more expansive than van Fraassen's. Does it also make it epistemically suspect?

The "pragmatic virtues" of a theory, van Fraassen writes, "are sometimes summarized as simplicity, sometimes as explanatoriness" (van Fraassen 1985, 285). He also calls them "superempirical virtues," but that is now just linguistic imperialism, and henceforth I am having none of it. The problem with pragmatic virtues, as he sees it, is that they give us no reasons for believing in the truth of virtuous theories.[6] If simplicity, for instance, is supposed to be a reason for belief, "we must ask which attribute of simplicity, of the theory or of the depicted world, is at issue." If the former, the thesis appears to be that the simpler products of our theorizing are "more likely to correspond to the facts," a claim that van Fraassen simply finds groundless. So perhaps the thesis rests on the conviction that *the world* is simple. But even if the world is simple, he insists, there must still be lots of simple false theories, and the true theories may not predominate among the simple ones. We can indeed use simplicity as a basis of theory choice, van Fraassen concludes, "but not on the grounds that this will increase our chances of metaphysical or even empirical success" (1985, 285–86).

What about explanatoriness? Although in fact I think that there is perhaps something (evolutionary) to be said on behalf of simplicity, my sympathies clearly lie here. Not to put too fine a point on it, I don't think that constructive empiricists—or, more precisely, that natural science as constructive empiricists understand it—can explain much of anything at all. But I am convinced that explaining things is an essential part of what natural science is for, what it is about, and what it is up to. Van Fraassen demurs, but he evidently agrees, at least, that this is where the dialectic ultimately leads. "In philosophical practice, the dividing line between [constructive] empiricists and others does indeed appear in connection with explanation. . . . When it seems that two theories fit the phenomena equally well, what criteria shall we use to choose between them? The answer will have to cite properties and relations going beyond considerations of empirical adequacy. Here explanation takes center stage" (van Fraassen 1985, 286). Since such considerations, "beyond evidence directly bearing on empirical adequacy," in fact do play a role in theory assessment and acceptance, he concludes, we face a dilemma: "On the one hand, we can profess to have reasons for believing these other considerations to yield reliable indications of truth. . . . On the other hand, we can attempt to remain [constructive] empiricists and conclude that these extra considerations which legitimately bear on theory acceptance are not legitimate reasons for belief. This includes the corollary that acceptance is not to be identified with belief" (van Fraassen 1985, 287).

Now van Fraassen concedes that the explanatory reach of a theory is indeed a virtue, that is, that it can be a reason for accepting the theory.

> [Ceteris paribus] theories are better if they are more explanatory, which in turn requires them to be more informative. But there are other reasons, besides the weight given to explanation, to conclude that, ceteris paribus, the better theories are the more informative. We go to science to have our questions answered about the empirical world, for many purposes—not just explanation, but practical control, mere factual curiosity ("thirst for knowledge," if you like), suggestions for new directions of research, and perhaps others. (van Fraassen 1985, 280)

But, he argues, precisely because explanatory power varies directly with the informativeness of a theory, it cannot serve as a reliable indicator of truth. That is why acceptance is not belief.

> At the same time, we worry that the information given by the theory may be false.[7] So we also compare theories with an eye to the possibility of falsehood, their credibility. However this comparison is carried out, and whether the criteria are objective or context-dependent or subjective, *credibility varies inversely with informativeness*. This is most obviously so in the paradigm case in which one theory is an extension of another: clearly the extension has more ways of being false. So our two models of evaluation are in radical tension, conflict—as I put it [elsewhere[8]]:

informational virtues are not confirmational virtues. Indeed, the two desiderata cannot be jointly maximized. They conflict; they detract from each other.

If some reasons for acceptance are not reasons for belief, then acceptance is not belief. And indeed some reasons for acceptance hinge crucially on the audacity and informativeness of the theory. So acceptance is not belief. (van Fraassen 1985, 280–81)

Now I would indeed agree that we compare theories with respect to their credibility—and that brings us to the crux of the matter. One of the basic convictions on which my disagreements with van Fraassen rests is that the only way that one could do this, even in principle, is by comparing theories with respect to their acceptability. Credibility varies directly with acceptability because reasons for acceptance are reasons for belief.

The second horn of van Fraassen's dilemma is constructive empiricism, which I reject. That suggests that I must accept the first horn. Do I then "profess to have reasons for believing" that explanatory considerations yield "reliable indications of truth"? Not as van Fraassen understands the claim.

My view is that explanatory considerations—what van Fraassen calls "reasons for acceptance"—are reasons for *belief*, and, of course, I have no quarrel with the platitude that "to believe is to believe true," that is, as Peirce put it, "we think each one of our beliefs to be true, and, indeed, it is a mere tautology to say so."[9] There is nothing troublesome about such a trivial immanent conception of "truth." The Tarski biconditionals satisfactorily fix its scope, and it can generally be interpreted along "minimalist" and "deflationary" lines, for example, "disquotationally" or "prosententially." But for just this reason, it is not the sort of thing for which there can be "reliable indications." For that to make sense, we need an epistemically transcendent conception of "truth."

Van Fraassen evidently thinks of truth in such epistemically transcendent terms, as a "metaphysical" relation of "correspondence with the facts" or "correctly representing the world." But I do not see how a constructive empiricist can make sense of such a notion, inter alia, because I cannot see what a constructive empiricist could mean by "the facts" or "the world." And, to return finally to my point of entry into this dialectic, these same considerations also apply to Longino.

It won't do to say only that a theory (a claim, a hypothesis, etc.) is true in the epistemically transcendent sense just in case it corresponds to the facts or correctly describes the world or says how the world is, for that is only to endorse the biconditional

TC. A theory T is true *if and only if* it corresponds to the facts (correctly describes the world, says how the world is).

But what such a biconditional tells us is only that any considerations which entitle us to either side of it will be considerations which entitle us to the other. It does not tell us what those considerations are—or even if there are any such considerations at all. That's the moral of Wittgenstein's elegant example (Wittgenstein 1953, §§350–51):

W. It's five o'clock on the surface of the sun *if and only if* it's five o'clock here and the same time on the surface of the sun.

In other words, we can know that two concepts must have the same conditions of application *if any* without knowing whether they *do* have any empirical conditions of application.[10]

As far as I can see, constructive empiricism implies that epistemically transcendent concepts of "the world" and "the facts" can't have any empirical conditions of application. More precisely, the claim that there is a world which our theories might or might not correctly describe (that there are facts to which our theories might or might not correspond) is at best a claim that, in accordance with constructive empiricist principles, we could have reasons to accept, but not one that we could have reasons to believe. But if we don't believe that there is such a world (that there are such facts), then we have no reason to distinguish between reasons for acceptance and reasons for belief. That makes constructive empiricism, if not actually internally incoherent, at least groundless.

Like any good Peircean pragmatist, I understand such notions as "the world" and "the facts," and hence the notion of truth operative in the biconditional TC, in terms of the practices of scientific inquiry. Peirce's own formulation of their relationships is both notorious and problematic: "Different minds may set out with the most antagonistic views, but the progress of [scientific] investigation carries them by a force outside of themselves to one and the same conclusion. . . . The opinion which is fated to be ultimately agreed to by all who investigate, is what we mean by the truth, and the object represented in this opinion is the real."[11] Here, however, I want to bracket considerations regarding any such supposed "convergence."[12] In the same vein, Peirce also claims that the cardinal virtue of sound scientific inquiry is that it invokes a method "by which our beliefs may be determined by nothing human, but by some external permanency—by something upon which our thinking has no effect."[13] And this is more useful.

The idea that I want to take from Peirce here is that what a scientific realist needs isn't an epistemically transcendent notion of truth, but rather an empirically acceptable conception of objectivity. That is what is fundamentally correct about van Fraassen's insistence that the crux of empiricism consists in the conviction that our beliefs should be responsive only to experience, for what we experience is not a matter of arbitrary choice or decision. The perceptual judgments that embody our observations are spontaneous. We don't reach them by cognitive processes lying within our deliberative control; we find ourselves with them. Perceptual judgments have a unique sort of de facto nonprocedural immediacy. That makes them evidence or data, but it does not make them infallible.

From the epistemological perspective, that is, a spontaneous perceptual judgment about how things are is no less an explanatory hypothesis than is an explicit theoretical posit. As I put it elsewhere, such a spontaneous perceptual judgment expresses "a hypothesis regarding the character of the item which, *qua* stimulus, has (causally) *evoked* it."[14] On this point, Peirce is especially clear: "Abductive inference shades into perceptual judgment without any sharp line of demarcation between them; or, in other words, our first premises, the perceptual judgments, are to be regarded as an extreme case of abductive inferences. . . ."[15]

But if this is right, then there is no in principle epistemological difference between what van Fraassen calls "empirical" and "pragmatic" virtues and, hence, none between observation and theory. That is the deepest reason why constructive empiricism ultimately cannot be sustained. In the last analysis, the acceptability of specific spontaneous perceptual judgments regarding observable things turns on considerations of overall explanatory coherence just as much as the acceptability of specific inferentially based theoretical claims regarding unobservable things does. Both sorts of claims are consequently equally intersubjectively defeasible in the light of further experiences, and that is enough to secure their objectivity, which is all that a scientific realist needs.

At least, it's all that my sort of scientific realist needs. Not everyone who calls himself a "scientific realist" will be satisfied with this, of course, and there is also a good deal more to be said about both van Fraassen's constructive empiricism and Longino's conception of science as social knowledge. But there's a definite moral to be extracted from what I have argued here, which I would insist needs to be respected in any discussion of scientific realism: We first need to get our general epistemology in order, and we cannot do so if we simply take for granted such notions as "truth," "reference," and "the facts" or "the world."

NOTES

1. This is not an entirely felicitous terminology, but it will do for setting up a rough initial contrast.

2. "To say of what is that it is not, or of what is not that it is, is false, while to say of what is that it is, or of what is not that it is not, is true" (*Metaphysics* 101 1b25).

3. I discuss the special case of perceptual observation below. For the special case of theory succession in mathematical physics, see Rosenberg (1980a, 1988).

4. Again: "I identify empiricism with the epistemological thesis that experience is the sole legitimate source of information about the world" (van Fraassen 1985, 286).

5. "The human organism is, from the point of view of physics, a certain kind of measuring apparatus. As such it has certain inherent limitations—which will be described in detail in the final physics and biology. It is these limitations to which the 'able' in 'observable' refers—our limitations, *qua* human beings" (van Fraassen 1980, 17).

6. As the reader will recall, this was precisely Longino's complaint as well.

7. The words are carefully chosen. On van Fraassen's account, we recall, "empiricism" is identified with "the epistemological thesis that experience is the sole legitimate source of information about the world" (van Fraassen 1985, 286). In light of his interpretation of "experience," then, the *information* given by a theory will consist in its observable consequences. The worries about truth mentioned here, that is, are always and only worries about the potential outcomes of possible measurements and observations. And regarding *them*, presumably, there can't be any question about whether what we believe does or doesn't "correspond to the facts." "The facts" just *are* what we observe. But now we have clearly wandered into contentious epistemological territory, for "the Given," i.e., classical foundationalism with its "self-warranting beliefs," is lurking just around the next dialectical corner.

8. At a symposium on scientific explanation with Wesley Salmon and Clark Glymour, about which van Fraassen gives no further information in the present text (van Fraassen 1985, 280).

9. From "The Fixation of Belief," in Peirce (1935, vol. 5, 375).

10. Another example: "X is a square circle *if and only if* X is an equilateral rectangle all of whose points are equidistant from a single center point." For more about the limitations of such biconditionals, see chapter 4 of Rosenberg (1980b).

11. From "How to Make Our Ideas Clear," in Peirce (1935, vol. 5, 407).

12. I explicitly discuss and assess Peirce's remark in previous work; see Rosenberg (2002, chap. 6). The essays cited in note 5 above are also apposite here.

13. From "The Fixation of Belief," in Peirce (1935, vol. 5, 384).

14. See Rosenberg (2002, 240). These ideas are worked out and defended in more detail there.

15. From Peirce (1935, vol. 5, 181).

REFERENCES

Frank, Philipp G. 1956. "The Variety of Reasons for the Acceptance of Scientific Theories," in *The Validation of Scientific Theories,* ed. P. G. Frank, 3–18. Boston: Beacon.

Longino, Helen. 1990. *Science as Social Knowledge*. Princeton: Princeton University Press.

Peirce, Charles Sanders. 1935. *Collected Papers of Charles Sanders Peirce*, vol. 5–6, ed. Charles Hartshorne and Paul Weiss. Cambridge, MA: Harvard University Press.

Rorty, Richard. 1991. *Objectivity, Relativism, and Truth*. Philosophical Papers, vol. 1. Cambridge: Cambridge University Press.

Rosenberg, Jay F. 1980a. "Coupling, Retheoretization, and the Correspondence Principle." *Synthese* 45:351–85.

———. 1980b. *One World and Our Knowledge of It*. Dordrecht: D. Reidel Publishing Co.

———. 1988. "Comparing the Incommensurable: Another Look at Convergent Realism." *Philosophical Studies* 54:163–93.

———. 1993. "Raiders of the Lost Distinction: Richard Rorty and the Search for the Last Dichotomy." *Philosophy and Phenomenological Research* 53:195–214.

———. 2002. *Thinking About Knowing*. Oxford: Oxford University Press.

Sellars, Wilfrid. 1963. "Empiricism and the Philosophy of Mind." In *Science, Perception and Reality*, 127–96. Atascadero, CA: Ridgeview Publishing, 1991.

van Fraassen, Bas C. 1980. *The Scientific Image*. Oxford: Clarendon Press.

———. 1984. "To Save the Phenomena." In *Scientific Realism*, ed. Jarrett Leplin. Berkeley, CA: University of California Press.

———. 1985. "Empiricism in the Philosophy of Science." In *Images of Science: Essays on Realism and Empiricism*, ed. Paul Churchland and Clifford Hooker. Chicago: University of Chicago Press.

Wittgenstein, Ludwig. 1953. *Philosophical Investigations*. Trans. G. E. M. Anscombe. London: Macmillan.

PART II

THE DEMANDS OF SOCIETY ON SCIENCE ❖ Socially Robust Knowledge and Expertise

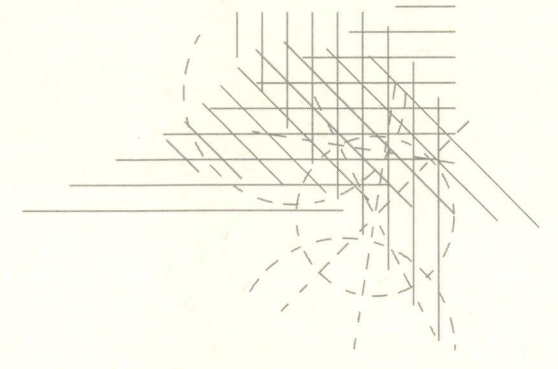

6

HOW ROBUST IS "SOCIALLY ROBUST KNOWLEDGE"?

PETER WEINGART
Bielefeld University

T HE ACADEMIC DEBATE ON THE DEMOCRATIZATION of expertise has reached the level of public politics. The U.S. National Research Council, in its study *Understanding Risk* (Stern and Fineberg 1996), has suggested "collaborative analysis" as a method adding deliberation to risk analysis and risk evaluation, thus opening advisory processes to broader participation. The British House of Lords Select Committee on Science and Technology, reacting to the devastating loss of credibility of expertise after the BSE (mad cow disease) crisis, published the report *Science and Society* (2000). Finally, and perhaps most prominently, the European Union, in a white paper on democratic governance (2001) produced in connection with a working group on "Democratizing Expertise," announced guidelines "on the collection and use of expert advice in the Commission to provide for the accountability, plurality and integrity of the expertise used" (5). It is justified to speak of a discourse that extends from academic discussions on the challenges of postnormal science and a new mode of knowledge production to public debates and declarations on public engagement with science and technology. A host of concepts reflect the change in perception that is taking place. Some of these have diffused into the public realm, while others are still confined to their academic origins. They all communicate the need to somehow bridge the lines between politics, power, and science, and truth. And they all proclaim the new regime of accountability. "Accountability" suggests that scientific experts are held responsible to political practitioners, and beyond, to their constituencies, that is, the public. "Quality" and "transparency" of scientific advice echo the same expectation voiced, in this case, by the European Union (1997). "Participatory technology assessment" (pTA) has become a movement that has led to a variety of experiments with consensus conferences, round tables, and similar devices of deliberation that bring together scientific experts and laymen (Abels and Bora 2004; Joss and Durant 1995).

All these terms allude to organizational devices that are intended to achieve accountability of scientific knowledge production, and perhaps even the democratic shaping of knowledge and technologies as they develop. In fact, the entire discourse is focused on the basic dilemma between democratic legitimacy by representation (vote) and the legitimacy conveyed by rationality on the basis of certified knowledge. The dilemma, of course, is old, but the discourse produces ever new answers, both terminological and institutional. The new term "socially robust knowledge" (SRK) refers to the underlying epistemological issue: how to accommodate democratic procedures of representation and decision by compromise, on the one hand, and the credibility, reliability, and quality of scientific knowledge claims, on the other (Nowotny, Scott, and Gibbons 2001). It is a cornerstone in the discourse insofar as it points to the nature of scientific knowledge proper rather than to institutional mechanisms with which to resolve the dilemma.

The concept of socially robust knowledge and its discursive career are well suited to reveal the underlying motives and conditions of the discourse on the democratization of expertise. Rather than adding yet another think piece on how democratization can be achieved and what the obstacles are, I intend to take a metaperspective and focus on the discourse itself.

Socially Robust Knowledge as a Rhetorical Device

"Socially robust knowledge" is a central element in Nowotny, Scott, and Gibbons's notion of Mode 2 science (2001). It is left open whether it is a descriptive term or a normative one, but it is clear that it is something we do not yet have. We are not told what exactly socially robust knowledge is. The first mention of the term in the book comes as an aside: "The reliability of scientific knowledge needs to be complemented and strengthened by becoming also socially robust. Hence, context-sensitivity must be heightened and its awareness must be spread. *The necessary changes pertain to the ways in which problems are perceived, defined, and prioritized*, which has implications for the ways in which scientific activities are organized" (Nowotny, Scott, and Gibbons 2001, 117, my italics). One can infer from this that socially robust knowledge has to do with problem perception, definition, and prioritizing and is also a part of contextualization.

Since there is an entire chapter called "The Context Speaks Back," one can hope to find more information on the nature of that process. The authors describe how science has shaped society ("spoken to it") and how now society "speaks back": "As a result, basic science has been *de facto* reconfigured in the context of the knowledge-based economy. The other half of the story concerns what we mean by contextualization of science" (53).

Unfortunately, the answer to this important question is hard to come by. Next we read that "a Mode 2 society generates the conditions in which society is able to 'speak back' to science. Contextualization is invading the private world of science, *penetrating to its epistemological roots* as well as its everyday practices, because it influences the conditions under which 'objectivity' arises and how its reliability is assessed" (55, my italics). Although the meaning of contextualization remains vague to this point, the authors rightly feel compelled to state that "it is necessary to demonstrate that contextualized knowledge is at least as objective as uncontextualized knowledge—*albeit in a different sense*" (55, my italics).

Next we are given two meanings of contextualization: "by pointing to shifts in research agendas and how research priorities are set, and describing how the policies of research councils and other funding agencies are articulated and directed towards certain objectives"; and "a second and deeper meaning which relates to our conceptions of how science 'really' works" (56). The authors want to scrutinize the second, "socio-epistemological meaning of contextualization" in order to understand "how the first affects the second" (56). After a discussion of shifting boundaries, we are told once again that "it has become necessary to explore the extent to which and in what ways these processes are affecting the core of scientific knowledge production. Is there a hard epistemological core underlying scientific knowledge production which cannot be changed without destroying what makes science work?" (58).

A few pages further on, following the discussion of the problems of peer review, the authors return to asking a question: "Why is there such resistance to admitting that the result—the commonly accepted reality as defined by science—is also open to social shaping, to cultural meanings, to integration into a life-world which science makes little attempt to explain?" (62). At the end of the chapter we are still without an answer, but we get the same normative claim again: "If scientists would openly acknowledge these perceived threats, it might be possible to develop another model of knowledge production, in which knowledge becomes *socially more robust*" (64, my italics). In addition, we are presented with another vacuous meaning of "contextualization," but while a few paragraphs earlier it was still an open question what effects contextualization would have *if* it happened, now the authors claim that it *has* already happened: "In historical terms it is clear that contextualization has surreptitiously crept into what was once held to be the inner core of science, while science's more outwards-oriented parts have actively and openly embraced contextualization" (64).

Alas, what we have just learned already exists is yet but a dream! Depending on a "political decision based on cultural considerations" yet to be made, "the actual practice of science . . . might be set free to explore different contexts and perhaps evolve in different directions. The research process would . . . be seen . . . more as a comprehensive, socially embedded process. . . . It is in this sense that we talk of the contextualization of science, as an enlargement of its scope and enrichment of its potential" (65).

Even though we still do not know precisely what contextualization (and therefore socially robust knowledge) means and whether it exists already or not, we are led to the next question: "How does it happen?" (chap. 7), that is, the question of "how contextualized knowledge is produced and which form it takes" (96). The answer has to do with the move of science from a socially segregated activity to an integrated one. This diagnosis is based on the unquestioned fact that universities educate for a broad labor market way beyond the reproduction of academia, and that scientific knowledge is in broad demand. A "second answer" is the "increasing prevalence and importance of uncertainty" (113). A little further in the text the authors have another insight: "Control from within," more elaborate peer review, formal quality control, and so on are "tiny cracks in the fabric of knowledge production through which contextualization enters" (115). However, the authors do insist that autonomy and independence of science have to be preserved.

Although at the end of that chapter we are still left in the dark about what contextualization means (remember, we want to know because it is an important ingredient of socially robust knowledge), we are at least given another example of what beautiful prose can achieve: "Contextualization . . . suggests a spectrum of complex interactions between potential and use, constraints and stimulants" (120). Not surprisingly, where there is a "spectrum" there is weakly and strongly contextualized knowledge, as well as examples of "middle-ground contextualization" (120).

Up to this point we know the following about socially robust knowledge and contextualization: It has to do with the priority setting and funding programs of research councils and other funding agencies. Scientists are apparently reluctant to acknowledge that these have changed, but if they did, science could develop its potential more fully. How far the contextualization of science has progressed is unclear, but while it is considered a good thing, it is not supposed to adversely affect the so-called epistemological core or the autonomy of science.

Chapter 11, entitled "From Reliable Knowledge to SRK," promises to give a direct answer as to the nature of socially robust knowledge (166). Here we learn:

> the more strongly contextualized a scientific field . . . the more socially robust is the knowledge it is likely to produce. What does that mean in practice? First, social robustness is a relational, not a relativistic or (still less) absolute idea. . . . Next, social robustness describes a process that, in due course, may reach a certain stability. Third, there is a fine but important distinction to be drawn between the robustness (of the knowledge) and its acceptability. . . . Of course, the two are connected—but social robustness . . . is prospective. . . . Fourth, robustness is produced when research has been infiltrated and improved by social knowledge. Fifth, and last, socially robust knowledge has a strongly empirical dimension. (167)

First, the authors turn to "reliability" as a major epistemic value. In their view, it should not be defined exclusively in terms of replicability but as validity outside "sterile spaces" such as laboratories (168). Then they enter into a lengthy discussion about the scope of consensus in view of specialization and the relationship between consensibility and consensuality pointing to different potential "experts" and practitioners. Given the expansion of communities of "stakeholders," "is it still possible to produce reliable knowledge?" (174).[1] This question is so important to them that they repeat it three times and then insist that "there is no suggestion that consensuality, like consensibility cannot be practised in such a context" (176; see also 173–75).

Having come this far, two issues still remain on the authors' agenda: "the potential for replication and the impact of secrecy" (176). Why these issues are selected is unclear, but the gist of the argument is that secrecy is counterproductive even in the Mode 2 setting, and replicability is no problem because it is not achieved anyway, either in Mode 1 experimental science or in Mode 2 (176). Finally, concluding this development, the authors have convinced themselves that "reliable knowledge, as validated in its disciplinary context, is no longer sufficient or self-referential. Instead it is endlessly challenged, and often fiercely contested, by a much larger potential community" (177). Then it is added, in good circular fashion, that "reliable knowledge—to remain reliable—has also to be socially robust knowledge" (178).

Toward the end of the book we are taken to the sanctum, the "epistemological core." While we still do not know what socially robust knowledge is, the authors insist that it does not violate the epistemological core, that autonomy and independence of science remain intact, and that scientific knowledge remains reliable. The plot thickens; suspense becomes almost unbearable. Here we expect to encounter the crucial argument. How is the contradiction between socially robust knowledge and the epistemological core resolved? "The most radical part of our argument, and therefore the most difficult to accept, is that the co-evolutionary changes . . . bundled together under the convenient label 'Mode-2 society' . . . have made it necessary not only to re-conceptualize the reliability of knowledge but also to question its epistemological foundations" (178). The solution to the riddle is simpler than anyone would have guessed: "the epistemological core is empty—or, alternatively and perhaps more accurately, crowded and heterogeneous." Their argument "therefore is that a more nuanced, and sociologically sensitive, account of epistemology is needed" (179).

To make a long story short about the "ambivalence of novelty" and the "decline of cognitive authority," this badly needed "sociologically sensitive epistemology" should "incorporate the 'soft' individual, social and cultural visions of science as well as the 'hard' body of its knowledge" (198). Thus the problem that seemed so insurmountable at the beginning has been successfully restated. Its dissolution into a harmonious accord of scientific method, new knowledge, and its social embedding is already caressing our senses: "the process of contextualization moves science beyond merely reliable knowledge to the production of socially more robust knowledge" (246).

To summarize, social robustness is, above all, a property that scientific knowledge should achieve. This could happen if science were to open itself to the social context. The nature of "context" remains vague, but it can be inferred that it means social and political concerns, the values and interests of lay publics that are directly or indirectly affected by scientific knowledge. Their voices are supposed to be heard, at least in democratic regimes. The authors of SRK are more ambitious than many others in that they take the dilemma of power and truth to the epistemological level. What they call the "epistemological core" is never clearly defined, either, but it is associated with the autonomy of science and the reliability of scientific knowledge (58). At the end, however, they state their "radical answer" to the dilemma: the epistemological core "is empty" (179). (It is an interesting side issue that the social sciences are deemed able to provide the "sociologically sensitive epistemology" that the natural sciences are lacking.) The term "socially robust knowledge," then, encompasses the dilemma and suggests, in programmatic fashion, its solution. To be convincing, the epistemological basis of science must be taken seriously at the outset, only to be cast aside in the end. It should not escape attention that the very vagueness of the terms involved explain their popularity among scholars of Science, Technology, and Society (STS) and practitioners of science policy because such vagueness creates the illusion that the dilemma can be solved. If the concepts were more concrete and if they had a better empirical grounding, the message would be more disappointing.

Socially Robust Knowledge and the Participatory Turn

The concept of socially robust knowledge has received some attention from other scholars. Thus it is worthwhile to look at interpretations and uses of the term in writings addressing the same dilemma in order to see which role they attribute to it and if and how they configure it. One appropriate place, among others, for such an exercise is a special issue of *Minerva* in which several authors respond to the precursor of *Re-Thinking Science* (Gibbons et al. 1994). Among them is Sheila Jasanoff, whose work has been focused for a long time on the special features of "regulatory science," and who supports her views with rich empirical observations. In her reaction she regards the concept of socially robust knowledge as the authors' "solution to problems of conflict and uncertainty" as they are endemic to the "pipeline model" of the relationship between science and society (Jasanoff 2003a, 235). However, she continues, the problem is "how to institutionalize polycentric, interactive, and multipartite processes of knowledge-making within institutions that have worked for decades at keeping expert knowledge away from the vagaries of populism and politics. The question confronting the governance of science is how to bring knowledgeable publics into the front-end of scientific and technological production—a place from which they have historically been strictly excluded" (Jasanoff 2003a, 235).

The answers to this crucial question are varied. One indisputable observation is to point to the "participatory turn," examples of which were cited above, such as the initiatives of the U.S. Congress to "concede unchecked autonomy to the scientific community in the collection and interpretation of data," the European Union's commitment to "involving public in technically grounded decisions," or the various "experiments . . . such as citizen juries, consensus conferences and referenda" (quoted in Jasanoff 2003a, 236–37).

Jasanoff herself observes the practical problems connected with these procedural and organizational innovations, not least the danger that participation may become "an instrument to challenge scientific points on political grounds." She states that "participation alone . . . does not answer the problem of how to democratize technological societies," and that the issue is "*how* to promote more meaningful interaction among policy-makers, scientific experts, corporate producers, and the public" (237–38, my italics). Her suggestion is to complement the existing "technologies of hubris" as they are embodied in the predictive approaches "(e.g., risk assessment, cost benefit analysis, climate modeling)" with new "technologies of humility" (238, 240). These are considered to be "*social* technologies" that "would give combined attention to substance and process, and stress deliberation as well as analysis," or, in other words, would "seek to integrate the 'can do' orientation of science and engineering with the 'should do' questions of ethical and political analysis" (243, my italics).

A sympathetic reading of Jasanoff's response to Gibbons et al. will readily acknowledge her analytical precision as well as the succinctness of some new key concepts. She terms the amalgamation and disciplining of scattered and private knowledge "civic epistemology." As a crucial component of the "technology of humility," the term conveys a meaning somewhat similar to socially robust knowledge. (Incidentally, she, too, sees the social sciences in the role of providing the substantive basis for these "technologies.") Also, she probes further into the meaning of socially robust knowledge by identifying "focal points" around which technologies of humility are to be developed: framing, vulnerability, distribution, and learning (240).

The difficulty, if not impossibility, of achieving any real progress (as opposed to mere conceptual innovation) is illustrated by another variant of bridging the expert/lay divide. Looking at the sources of legitimacy of experts in different political cultures (in the United States, the United Kingdom, and Germany), Jasanoff sees the reason for the foreclosure of "continuous dialogue between expert and critical lay judgment" in the imperfect framing of the problem of expertise in democratic societies (Jasanoff 2005). She therefore suggests it would be better "to recast the role of experts in terms that better lend themselves to political critique," that is, to subject expert decision making to "notions of delegation and representation." In such a framework, experts will act "not only in furtherance of technical rationality, but also on behalf of their public constituencies" (Jasanoff 2005, 222). However, she insists that "*equally,* citizens need to recognize that governmental experts are there to make judgments on behalf of the common good rather than as spokespersons for the impersonal and unquestionable authority of science." The resulting expectation is "that a fullfledged political accountability—looking not only inward to specialist peers but also outward to engaged publics—*must* become integral to the practices of expert deliberation" (Jasanoff 2005, 222, my italics; cf. also Jasanoff 2003b).

These examples are sufficient. They are from one of the most sophisticated observers of expert advice to policy making, and thus there are no hidden secrets or surprises of institutional innovations to be expected. Two conclusions can be drawn from the above. First, the discourse on the democratization of expertise is clearly a normative one. Second, a look at close range shows that for all their conceptual ingenuity, the analyses end up in a continuous restating of the dilemma. Why is that so? What keeps the discourse going despite the impasse?

Discursive Strategies and the Limits to Transgression

The dilemma of power and truth cannot be resolved and can therefore only be processed by way of discourse (Maasen and Weingart 2005). This is the common cause of the discussion's impasse and its normative nature. Both sides of the dilemma are deeply ingrained in our culture and have been theoretically conceptualized as their functional differentiation (Luhmann 1990). This means that as long as modern societies are organized on the principle of functional differentiation, there will be no room for a blurring of boundaries. The very fact that a sizable array of metaphors continue to be invented that suggest management or mediation of the boundary ("trading zones" giving rise to "contact languages" and "transaction spaces," "transgressivity," etc. [Nowotny, Scott, and Gibbons 2001, 145]) points toward the ironclad existence of that boundary.

Does this mean that the empirical descriptions of the changes conveyed with these metaphors are wrong, and does it mean that nothing can be done about the institutional arrangements of expert advice to policy making? The answer is no to both questions! The issue is rather what can be expected.

In terms of discursive strategy, two options are employed that seem to suggest that the dilemma can somehow be resolved. One is to argue that scientific knowledge is not truth-bound as much as it is supposed and believed to be. Keeping to the symmetry of the dilemma, the other option is to argue that political decision making is not entirely power-oriented but is rationality-oriented as well. Curiously, the first option receives much more attention than the latter from analysts of the interface between experts and policy makers. This imbalance of attention may itself be taken as an indicator of the preoccupation with scientific expertise and the privileged role of experts and their knowledge as somehow foreign to the ideals of democratic representation. The primary concern in the discourse is the invasion of "truth" (that is, knowledge held by few) into democratic decision making. The opposite option, that is, the invasion of political considerations into the search for knowledge, receives comparatively less attention. This has not always been the case. Concerned by instrumentalizations of science by authoritarian fascist and socialist governments, writers such as Robert Merton, Karl Popper, and Michael Polanyi were primarily concerned with the politicization of science. Thus, it matters who observes, and when.

The former part of the discourse is determined by sociologists and science studies scholars who, given their professional preoccupation with the highly specialized study of science, argue against the view of science as universal, certified, and unequivocal knowledge, and experts as neutral, politically disinterested arbiters of that knowledge. A plethora of publications is devoted to demonstrating, either by theoretical speculation or by empirical study, that experts play a much more ambiguous role, that they are not neutral, that they do not always restrict their advice to the realm of their expertise, and likewise that their knowledge is rarely certain, nor is it consensual, and that it is selective with respect to the questions asked, and so forth (Jasanoff 1990; Weingart 2001; Maasen and Weingart 2005). In addition, there is an equally elaborate literature on democratic, participatory technology assessment based in part on actual experiences with round tables, consensus conferences, open forums, and similar devices intended to involve the lay public in decisions on the design and implementation of new technologies (Joss and Durant 1995; Renn, Webler, and Wiedemann 1995; Abels and Bora 2004).

The gist of this part of the discourse seems to be that the apparent dilemma of power and truth is softened on the knowledge side: If the knowledge conveyed by experts is not the inimitable truth that is only accessible to a small elite, but rather is closer to negotiable opinion, then it is easier for the lay public to be successfully involved. Furthermore, if the role of experts is inherently political in the very context of giving advice, it is not legitimate for them to have privileged access to power. The demand that they be accountable is irrefutable.

The persistence of the dilemma in spite of this dismantling of the epistemic core of scientific expertise is perhaps best demonstrated by calling upon a witness from the "sociopolitical" side to comment on the other side of the discourse. Jasanoff diagnoses "the technocratic turn in US policy" characterized by its reference "back to an outmoded view of expertise as certified, elite knowledge and judgment" (Jasanoff 2003b, 161). At the same time, however, she observes the "slanting of expertise toward particular ends" happening in the (politically mandated) restructuring of the scientific advisory committee of the U.S. Department of Health and Human Services (HHS) and considers this a "different but equally thorny puzzle" (Jasanoff 2003b, 161). In effect, the George W. Bush administration kept the advisory committee intact but replaced members with handpicked choices. This may be read either way: as a technocratic turn of politics insofar as the role of experts as conveyors of certified knowledge is upheld, and as a political instrumentalization of scientific expertise insofar as the experts are chosen according to political criteria (such as the experts' ideological allegiance). This is a "thorny puzzle" only if primary attention is given to the administration's use of experts as sources of reliable knowledge. The other, neglected side is that experts are used as sources of legitimation, but only if their views correspond with those of the politicians that ask for their advice.

Curiously enough, policy makers seem to have no qualms about holding on to an "outmoded view of expertise," asking for sound and reliable advice, and instrumentalizing that advice for political ends at the same time. The supposed technocratic turn, as exemplified in the call for sound science, quality control of expert advice, and rosters of recognized experts (as intended by the EU), may in part be a reaction to the deterioration of scientific advice by politicization and the realization that politically instrumentalized experts do not deliver the credibility and, thus, legitimacy that is expected from their advice. The strength of this line of reasoning should not be underestimated, because it reflects the dominant view that policy makers and even the public have of science and experts, as any reading of media reports on incidents of political instrumentalization of scientific advice will readily show (Weingart 2005, chap. 5).

The gist of this less prominent part of the discourse seems to be that the dilemma is softened on the policy-making side. If policy makers cannot rely on democratic delegation of power alone because they are faced with problems emanating from science and technology, they are likely to attempt to gain political control of the expertise they have to enlist without losing its functionality both in terms of its problem solving and its legitimating capacities.

One incident during Gerhard Schröder's reign as chancellor of Germany may illustrate this. The minister for Consumer Protection, Nutrition and Agriculture, a member of the Green Party, was unhappy with the Central Commission on Biological Safety (ZKBS) regarding its friendly position towards genetically modified plants and organisms. Instead of attacking the scientific advisors directly or dissolving the commission, the ministry changed its mission and structure. It added criteria such as "ethical nature" and the protection of consumers and conventional farming to the evaluation of risks of genetic engineering. In order to implement these criteria the commission was divided into two panels, one that retained the commission's previous competence for work on genetically modified products, the second responsible for their distribution. The latter has to apply the new criteria. Thus, on the one hand, the expertise was to be adapted to the values of the Greens, on the other hand, the scientific reputation of the old commission was not given up.

This softening of the dilemma of power and truth has carried the discourse far beyond its original rigid state as it was represented in Max Weber's classic decisionistic juxtaposition of politics and science. It has led to novel institutional arrangements that appear to be hybrid, and to new alignments of experts and laypeople in which knowledge, values, and interests are combined in transgressive modes. But does this mean that the dilemma has truly been resolved?

The irreducible concern about a completely technocratic politics, just as much as about a complete reduction of science to politics, suggests the contrary. The more far-reaching visions of the accountability of experts remain programmatic. The most advanced arrangements of public participation in science- and technology-related decision making remain carefully staged exercises, special events with no institutionalized connections to existing forms of democratic representation and decision making. They are looked upon with great suspicion by policy makers, who jealously guard their legitimate claims to power. There is no escape from the logic underlying the dilemma: Strengthening accountability and public participation in expert judgments ultimately compromises the reliability and legitimating power of the experts' specialized knowledge, and likewise, putting experts in the position of determining political agendas and decisions conflicts with democratic representation and legitimation of power. The expectations invested in the emergence of a socially robust knowledge cannot realistically reach beyond these limitations.

Are There Alternatives to Robust Knowledge?

The notion of socially robust knowledge adds yet another twist to the classic problem of how to control science and technology in a modern democratic society which is more dependent on science and technology than ever before and, thus, also more susceptible to their risks. This debate began in the 1960s, when damage to the environment resulting from technological progress first became an issue. The subsequent debate can be reconstructed as the successive design of mechanisms whereby the control is gradually moved upwards from intervening at the stage of implementation of knowledge to that of the production of new knowledge. Or, as it may be phrased, the controls are moved from an *ex post* to an *ex ante* stage of crucial knowledge production. The rationale for this is the often-lamented fact that once knowledge is available, little can be done to stop its diffusion and implementation. Thus one would have to control its production in the first place. However, the logic of this progression is flawed for at least two reasons. One is the implicit crossing of the boundary between the implementation and the production of knowledge. Whereas the former can be subjected to democratic control, that is not as self-evident with respect to the latter. At best, one can think of a (democratic) control of the political decisions that determine (temporary) priorities of public funding of research. Here, values to be respected and risks to be avoided can be taken into consideration, as in fact happens already. Even these decisions are subject to repeal in the light of new knowledge. What would come closest to generating another kind of knowledge would be to strengthen mechanisms that would give a voice in implementing new knowledge to those who are affected by it, that is, to make their acceptance a criterion of implementation. But that points to a second reason why the concept is flawed: the recourse to "society." Marilyn Strathern has pointed to the problems of such an abstract notion of society, which clearly are not intended by the promoters of socially robust knowledge: "It produces the concept of 'science' (or technology, or academia) in contradistinction to itself; this de-socializes 'science and technology' as somehow less part of society than arts and humanities. . . . But above all, the invocation of 'society' summons the fragility of measurement: What will count as 'society'?" (Strathern 2005, 476). Indeed, this question is not posed, and the implication is disquieting. Either a diffuse, unspecified public is invoked (strikingly similar to that of the campaigns for "public understanding of science"), or lobby interests that are not legitimated by popular vote will prevail in this unstructured public because their backers have the capital and the manpower.

The only escape from the difficulties of imagining institutional mechanisms of representing society vis-à-vis scientists would seem to be the implantation of a "socially responsible" conscience in the individual scientist's mind. Since that option is also not available, notwithstanding the relegation of many biomedical problems to the responsibility of scientists and to ethics committees, the concept of socially robust knowledge actually marks a turning point in the debate over the control of science and technology. Rather than carrying controls, however democratic, into the realm of knowledge production, the focus must be on institutional mechanisms that process new knowledge and values and interests at the same time. Round tables, consensus conferences, ethics committees, and similar stagings are only a beginning, and it is important not to overburden them with misplaced expectations. Their function can only be to serve as exemplars of the discursive processing of science- and technology-related issues in the broader public. They cannot control the production of knowledge, nor can they be representative of social values and interests. Rather, they can contribute to a better understanding of the scientific issues on the side of the lay public, and a better understanding of the lay public's concerns with regard to scientific knowledge and its social implications. Thus, the issue is to provide for institutional frameworks that allow for an unrestricted and unbiased deliberation of new knowledge and new technological options in the light of existing values and interests, because in this process, all of them will change.

NOTE

1. "'Consensibility' is supposed to mean that each message should not be so obscure or ambiguous that the recipient is unable to give it whole-hearted assent or to offer well founded objections" (Nowotny, Scott, and Gibbons 2001, 170).

REFERENCES

Abels, Gabriele, and Alfons Bora. 2004. *Demokratische Technikbewertung*. Bielefeld: Verlag, transcript.

European Union, Commission of the European Communities. 1997. *Communication of the European Commission on Consumer Health and Safety*. Cited in Jasanoff 2003a, 236.

European Union, Commission of the European Communities. 2001. *European Governance, A White Paper*. Brussels.

Gibbons, Michael, et al. 1994. *The New Production of Knowledge. The Dynamics of Science and Research in Contemporary Societies*. London: Sage.

House of Lords, Select Committee on Science and Technology. 2000. *Science and Society*. Science and Technology, Third Report.

Jasanoff, Sheila. 1990. *The Fifth Branch: Science Advisers as Policymakers*. Cambridge, MA: Harvard University Press.

———. 2003a. "Technologies of Humility: Citizen Participation in Governing Science." *Minerva* 41:223–44.

———. 2003b. "(No?) Accounting for Expertise." *Science and Public Policy* 30 (3): 157–62.

———. 2005. "Judgment under Siege: The Three-Body Problem of Expert Legitimacy." In *Democratization of Expertise? Exploring Novel Forms of Scientific Advice in Political Decision-Making*, ed. Sabine Maasen and Peter Weingart, 209–24. Sociology of the Sciences. Dordrecht: Springer.

Joss, Simon, and John Durant, eds. 1995. *Public Participation in Science: The Role of Consensus Conferences in Europe*. London: Science Museum.

Luhmann, Niklas. 1990. *Die Wissenschaft der Gesellschaft*. Frankfurt: Suhrkamp.

Maasen, Sabine, and Peter Weingart. 2005. "What's New in Scientific Advice to Politics?" In *Democratization of Expertise? Exploring Novel Forms of Scientific Advice in Political Decision-Making*, ed. Sabine Maasen and Peter Weingart, 1–20. Sociology of the Sciences. Dordrecht: Springer.

Nowotny, Helga, Peter Scott, and Michael Gibbons. 2001. *Re-Thinking Science. Knowledge and the Public in an Age of Uncertainty*. Cambridge: Polity.

Renn, Ortwin, Thomas Webler, and Peter Wiedemann, eds. 1995. *Fairness and Competence in Citizen Participation*. Dordrecht: Kluwer.

Stern, Paul C., and Harvey V. Fineberg, eds. 1996. Committee on Risk Characterization, National Research Council, *Understanding Risk: Informing Decisions in a Democratic Society*. Washington, DC: The National Academies Press.

Strathern, Marilyn. 2005. "Robust Knowledge and Fragile Futures." In *Global Assemblages, Technology, Politics, and Ethics as Anthropological Problems*, ed. Aihwa Ong and Stephen J. Collier, 464–81. Oxford: Blackwell.

Weingart, Peter. 2001. *Die Stunde der Wahrheit?* Weilerswist: Velbrück.

———. 2005. Die Wissenschaft der Öffentlichkeit. Weilerswist: Velbrück.

7

IN DEFENSE OF SOME SWEEPING CLAIMS
ABOUT SOCIALLY ROBUST KNOWLEDGE

ROGER STRAND
University of Bergen

S CIENTIFIC KNOWLEDGE AND PRACTICE, AND THE relationships between science and society, are the subjects of a great diversity of scholarly studies. Not only do the classical disciplines such as history, philosophy, sociology, and anthropology study various aspects of the scientific enterprise, but recent decades have seen the rise of new academic fields with their own methodologies developed in the quest to understand science. These include Science, Technology, and Society Studies (STS); the Sociology of Scientific Knowledge (SSK); actor-network theory (ANT), and so forth. Moreover, anybody knowledgeable in the philosophy of science will know that this apparently well-defined subdiscipline is by no means a homogeneous field characterized by consensus about its proper methodologies, purposes, styles, or even object of study. The gap to be bridged between Karl Popper and Gaston Bachelard is considerable. Currently there is no smaller distance between the philosophies of, say, Philip Kitcher and Jerome Ravetz.

One response to this breathtaking diversity is to run into one's own pigeon-hole, to "professionalize," so to speak, and to consolidate one's own position. There is nothing very wrong with this response, unless it is exaggerated into intellectual isolation. There is always a potential for fertilization of thought across traditions and disciplines. One may look to the German and Scandinavian languages, in which the concept of *Wissenschaftstheorie* has shown its value as a device for mutual recognition and dialogue.[1]

Differences in position, style, and academic culture may prove useful to fuel debate and stimulate intellectual vitality. However, they may also lead to misunderstanding, impediment of dialogue, and mutual distrust. For instance, some of the literature of the 1970s on the sociology of scientific knowledge and its reception by more analytically inclined philosophers may give the impression of a quite hostile relationship between practitioners of sociology of scientific knowledge and Anglophone "mainstream" philosophy of science. To the present author (neither a proper philosopher nor a social scientist but a former biochemist retrained in *Wissenschaftstheorie*), it appears that an important academic task for people like myself is to contribute to the work of translation and explanation required to grease the machinery of contact and exchange between the mentioned traditions and disciplines. I shall try to defend some of the apparently exaggerated claims to be found in a particular strand of scholarly studies of science that we might call a kind of radical philosophy or social theory within a context of science policy and governance. Specifically, I shall focus on the idea of socially robust knowledge as proposed by Helga Nowotny, Peter Scott, and Michael Gibbons (2001) and the related idea of postnormal science introduced some years earlier by Silvio Funtowicz and Jerome Ravetz (1993). It is my experience, perhaps more in oral than in written discussion, that criticism of these authors often lies in what is perceived as the excessive boldness and generality of their claims, going beyond the argument and evidence they provide. I have called their claims "sweeping"; in the next section their nature will become clear. Being sympathetic to what I perceive as the objective behind the theories of socially robust knowledge and postnormal science, I shall here provide a defense to which the original authors would not necessarily subscribe.

Some Sweeping Claims about Socially Robust Knowledge

To my knowledge, the expression "socially robust knowledge" dates back to communications by Nowotny (1999) and Gibbons (1999). In *Re-thinking Science* (Nowotny, Scott, and Gibbons 2001) the underlying idea is presented at length, though no formal definition is provided. The authors state that a desirable property of knowledge is that "it remains valid outside these 'sterile spaces' created by experimental and theoretical science, a condition we have described as 'socially robust'" (168).

The idea of validity appears to be connected to a concept of stability in society: "[S]ocial robustness describes a process that, in due course, may reach a certain stability. . . . [S]ocial robustness, in an important sense, is prospective; it is capable of dealing with unknown and unforeseeable contexts. . . . However, this is a reversal of the traditional pattern of scientific working, which has been to restrict as far as possible the range of external factors, or contexts, which must be taken into account" (167). Although the authors never refer to the writings of the philosophers Silvio Funtowicz and Jerome Ravetz, one is reminded of their idea of a new type of problem solving called "post-normal science" (Funtowicz and Ravetz 1993):

> Whereas science was previously understood as steadily advancing in the certainty of our knowledge and control of the natural world, now science is seen as coping with many uncertainties in policy issues of risk and the environment. In response, new styles of scientific activity are being developed. The reductionist, analytical worldview which divides systems into ever smaller elements, studied by ever more esoteric specialism, is being replaced by a systematic, synthetic, and humanistic approach. (739)
>
> [It involves] the inclusion of an ever-growing set of legitimate participants in the process of quality assurance of the scientific inputs. (752)

Among the legitimate participants in the democratization of expertise (in the case of AIDS), Funtowicz and Ravetz mention groups such as sufferers, carers, journalists, ethicists, activists, and self-help groups.

If we are allowed to talk about such a thing as a political "establishment," we may first of all note that this establishment does not have to experience difficulties with accommodating the ideas presented in the quotations above. It is true that applications of scientific knowledge, in the form of technology and policy advice, have had indirect, unpredictable and sometimes negative consequences in society. It is also true that the unpredictability is due to the mismatch between the complexity of the world and the simplicity of the knowledge, either in terms of its form (simple laws) or in terms of the idealized character of its primary range of validity (the laboratory, under controlled conditions). Third, it is also easy to recognize the desire of being able to cope better with the problems arising in the interface between society and the application of science. Indeed, authors such as Nowotny, Scott, Gibbons, Funtowicz, and Ravetz have had some impact on policy in the European Union, for example in the Science and Society initiative in the union's Sixth Framework Programme. Increased public participation may improve public understanding of, trust in, and cooperation with science, and integration of the knowledge base of public policy and decision making into our common body of knowledge is undoubtedly desirable.

However, the mentioned authors underline that such an interpretation does not capture the entire message, which appears to include a radical reevaluation or redefinition of science: "[B]asic science has been de facto re-configured in the context of the knowledge-based economy" (Nowotny, Scott, and Gibbons 2001, 53). "Contextualization is invading the private world of science, penetrating to its epistemological roots as well as its everyday practices, because it influences the conditions under which 'objectivity' arises and how its reliability is assessed" (ibid., 54). In slightly less delicate prose, Funtowicz and Ravetz (1994) asserted: "In [post-normal] science, science is no longer imagined as delivering truth, and it receives a new organizing principle, that of quality. This is dynamic, systemic and pragmatic, and therefore requires a new methodology and social organization of work" (1994, 197–98).

The task of the remainder of this paper is to mediate between these statements and what I have experienced to be their reception among more analytically inclined philosophers of science. How can one claim that science has acquired a new organizing principle? Does the interaction with society really penetrate to the epistemological roots of science? Is truth replaced by pragmatic considerations of quality? How should one understand such claims?

Explaining the Context

Before entering into polemics, for instance as to whether the laws of physics are valid only inside "the sterile spaces" in which they can be experimentally verified, or about the exact content of the epistemological core of science, it is useful to obtain a more global understanding of the perspectives of Nowotny, Scott, and Gibbons and Funtowicz and Ravetz: which actual problems in the world do they have in mind? What do they believe to be at stake when they make their "sweeping" claims? In other words, we ought to do ordinary intra-, inter-, and con-textual hermeneutic work to understand (in the sense of *Verstehen*) what they try to say.

The texts to which I have been referring are not justly represented by the presented samples. In fact, large parts of the texts are devoted to the more or less detailed presentation and discussion of episodes or facts from recent history of science and technology. Nowotny, Scott, and Gibbons (2001), as well as the preceding Gibbons et al. (1994), typically discuss examples of new forms of collaboration between different kinds of experts, or experts and nonexperts, in industry and social planning. Funtowicz and Ravetz have presented and discussed examples of inadequate assessment and management of industrial, technological, and environmental risks.

By this observation I do not intend to limit the generality of their claims by saying that they are "merely" about the practical details of, say, risk assessment and management. That would be similar to reducing Plato's world of ideas to speculative comments about geometry, or Popper's criterion of falsification to an argument about Alfred Adler's individual psychology or Marxist or Freudian theories. A philosopher who does not know of Pythagorean mathematics or the intellectual debates in Vienna in the first decades of the twentieth century does run a serious risk of misunderstanding Plato's or Popper's writings, though.

Misinterpretation is an even greater risk in the strands of theory in which the writings on socially robust knowledge and postnormal science occur, because the importance of context is actually emphasized by these theoretical perspectives themselves. To take Nowotny, Scott, and Gibbons seriously on their own terms, we have to ask about the context of their claims. Fortunately, we may ask them directly. Alternatively, we may "ask" the texts, in which case the reply appears to be given by the concrete episodes, facts, and issues they provide. Furthermore, in order to understand the acclaim experienced by some of these texts, we may ask the sympathetic reader not only if he believes in the literal truth of the general claims to be found, but also if he himself finds himself confronted with real-world problems similar to those approached in the texts.

I have been studying problems that in some ways resemble the case studies of the texts under discussion, with features more or less well captured by the quoted claims. For some years, I studied what we called the in vivo/in vitro problem in biochemistry. This consists in the methodologically unhappy situation that biochemists want to learn about biological processes in living organisms, but frequently have to destroy or strongly perturb these organisms as a direct consequence of their measurement techniques. For instance, one needs to kill the animal to take out an organ; or study an artificial cell culture; or study the properties of a protein in a highly purified aqueous environment, although its native, physiological environment is anything but homogenous, pure, or simple. Many practical methodological choices must be made regarding isolation and purification procedures and measurement conditions, and they may directly affect the outcome of the measurement. Worse, there may be no (in vivo) point of reference to which one may calibrate one's methodological setup, because the true in vivo situation is not directly accessible; this was exactly why one opted for the indirect, in vitro approach (Strand, Fjelland, and Flatmark 1996; Strand 1999). Accordingly, there is a methodological underdetermination that in certain cases may lead to a qualitative uncertainty in the resulting knowledge that is difficult or impossible to control or quantify. In our view, this uncertainty has political implications for the application of biotechnology, for instance for the design of risk assessment and management of the release of genetically modified organisms (GMOs). Specifically, we have argued for the inadequacy of a management system based upon strong belief in current quantitative risk assessments of certain undesirable events with GMOs (Strand 2000, 2001).

A comparable example is provided by our recent study of the Norwegian management system for issuing permissions for maintenance dredging in harbors (Cañellas-Boltà 2004, Cañellas-Boltà, Strand, and Killie 2005). Maintenance dredging is a desirable if not necessary function in active harbors. However, at many locations, the sediments to be moved during dredging operations have been found to include high concentrations of toxic chemicals such as PCB (polychlorinated biphenyl) and TBT (tributyltin). Some measures can be taken to limit or avoid the negative environmental consequences of the resuspension or relocation of these materials but they may be costly, while their effectiveness may remain unknown in the individual case. According to Norwegian regulations, some sort of assessment of environmental risk and impact must be performed in the decision-making process in order to "get the facts straight" and make an environmentally sustainable and cost-effective decision. The facts are anything but straight, though. One may estimate the amount of toxic chemicals in the sediments and their transport during dredging, but their biological significance at a given location with its particular topography, microclimate, and ecology may vary from very high to virtually zero, and it will often have to remain unknown. This is also the case because the outcome may be seen to depend upon the exact nature of long-term human activity, including changes in the economic activities of the local communities. Still, the decision-making process is managed *as if* it is possible for scientific experts to produce a credible risk assessment upon which a "rational choice" may be made. Indeed, consultants are hired to produce the desired assessments.

Translating back into the theoretical discourses of socially robust knowledge and postnormal science, we may say that the problems here encountered are related to the danger of in vitro biochemical knowledge or inadequate environmental risk assessment misleading our decisions on the societal level. This is possibly a metaphorical formulation, since it may be argued that knowledge cannot mislead. It is people who mislead and can be misled. What happens in practice, though, is that the decision makers are faced with a lack of knowledge of the real system to which the decision applies, the "in vivo" ecosystem and society in its full complexity. About this system one typically will have only anecdotal or other low-quality information. Expert institutions with scientific credentials provide information of high precision but with uncertain relevance, since it is produced either in a simplified in vitro laboratory system or, in the case of an environmental risk assessment, by applying a large number of simplifying assumptions for which there may be very little evidence in the actual case.

In physics, one defines "precision" as the ability of a measurement to be consistently reproduced, while "accuracy" describes the proximity to the "true" value. We may reformulate our problem as follows: The decision-making system tends to place a lot of confidence in expert advice due to the high precision of their in vitro knowledge, but in fact this level of confidence is not justified, because there are good reasons to believe that the accuracy of the advice is unknown or even low. However, this reformulation captures only half of the story, and it does so in an instructive manner: If the problem only were one of low accuracy, the cries for a reevaluation or redefinition of science would be absurd. The appropriate reply would then have been simply a refinement and development of existing science, increasing the accuracy.

I assume that Nowotny, Scott, and Gibbons, and Funtowicz and Ravetz are not against the refinement and development of scientific measurement methods. Indeed, Ravetz warned quite early of the risk of loss of skills and quality in the transition of science from an elite activity to a mass phenomenon (Ravetz 1971). However, I take the call for socially robust knowledge and postnormal science to be based in a disbelief that "more of the same," that is, more refinement and development of the science we already have, will suffice in the face of the challenges posed by the necessity of a change into sustainable development.

This view is based in a theoretical belief that has arisen in many strands of scholarly studies of science ever since Francis Bacon, Descartes, and Galileo, namely that analysis, reduction of complexity, simplification, idealization, in whatever words we describe it, has been an important condition for scientific success. Galileo and Descartes imagined that analysis could be followed by synthesis to the extent that one could gain both complete precision and accuracy; Bacon envisaged a future in a New Atlantis. During the twentieth century, the suspicion has emerged, however, that the success of science in the modern world is also partly due to a move toward granting higher importance to things that science indeed does control. Thus, Husserl ([1954] 1999) argued that the success of Galilean science led to the construction of a worldview out of what was a method, "the geometrization of the life-world"; systems theory or indeed any postempiricist philosophy has argued that the observer has to be considered as part of the system and that it is impossible to perform the complete synthesis from the "point of nowhere"; and the scientific discoveries of chaos, fractals, self-organized criticality, and other processes far from equilibrium indicated how much may have been systematically left out of scientific inquiry on the default assumptions of equilibrium or linearity (Strand 2002). When science has been applied so successfully, it may accordingly be explained in some cases by the good match between a relatively simple part of reality and its representation (theoretical knowledge), or by more complicated mechanisms in which power is applied to change a part of reality so that it conforms with the knowledge. Hence the architect Friedensreich Hundertwasser (1997) saw the ubiquitous introduction of Cartesian shapes (straight lines, plane surfaces) as dangerous to our psyche and humanity. Latour (1988) sees Pasteur not only as a great scientist but also as an important social force in the transformation of French dairy production, agriculture, and society. Similarly, Georges Canguilhem ([1966] 1989) argues that the development of physiology leads to a redefinition of disease disregarding the subjective experience of the patient and in terms of the identification of physiological deviations from normalcy; the societal consequences of this were explored by his student Michel Foucault ([1963] 2000). And even without the aid of French scholars, we may note the role of scientific and technological development in the immense increase in consumption of materials and energy throughout the last century.

Hence we are left with a picture in which scientific and technological development played and plays a crucial role in the making of long, safe, and comfortable lives (for rich people in modern welfare states, above all), but also by the same dynamics contributes to produce alienation, exploitation, and unsustainability. This family of theories and insights was generally overlooked by logical positivism or empiricism and related philosophies of science, because they focused their studies on scientific theories, and in particular their logical properties. However, in order to explain the context of other types of scholars studying science, it is necessary to keep in mind this background of European intellectual development in the twentieth century, all leading to a decline in optimism on behalf of "Galilean" science and technological development.

Demythification, not Denunciation of Science

Still, from the fact that science did not bring us Utopia it does not follow that new and sustainable types of knowledge production really are possible. When reading texts about socially robust knowledge or postnormal science, one may perceive an ambiguity as to whether they are descriptive or normative: Are the authors describing actual changes in knowledge production, or do they describe how they would like it to be? Looking only at the form of isolated sentences or passages, we may remain puzzled: some of them are logically descriptive and others normative. Remembering the context, it becomes clear that the authors' project transcends a mere description of matters of fact. This is quite explicit in the case of Funtowicz and Ravetz, whose objective is to propose novel strategies for knowledge production, for the communication and management of uncertainty, and for decision making in the interface between science and society in general. I think there is a shift in the normative direction from Gibbons et al. (1994) to Nowotny, Scott, and Gibbons (2001), especially in the last chapter of the latter text. Still, the texts are rich in descriptions, that can be thought of as exemplary, in the sense that they are examples indicating where to look and what to do when undertaking the work proposed by the normative statement. For Funtowicz and Ravetz, the promising developments are to be found in new, untraditional, radical disciplines such as ecological economics (Funtowicz and Ravetz 1994). Nowotny, Scott, and Gibbons find interesting developments also within the capitalist/industrial system. In their view, the "linear model" that describes a direct path whereby the results of basic science are used in applied science, which leads to technological progress, simply is not correct, at least not anymore. Instead, the industry has itself developed more efficient ways of making use of expertise in transdisciplinary collaboration, or what famously was called "Mode 2" science (Gibbons et al. 1994). And in any case, these scholars have to be considered not only as observers but as conscious actors in the interface between science and society. As such actors, their objective might be to be creative and supportive of new ideas and attempts, rather than to perform descriptive or normative research. An important example is the activity generated by Silvio Funtowicz and his coworkers at the Joint Research Centre at Ispra, Italy.[2]

One banal reason for explaining all this is to avoid misinterpretations of these authors as "anti-science." The target of criticism is not science, scientific practice, or practicing scientists. The target is a set of myths about science that serve to uphold an inappropriate confidence in expert advice in policy making under uncertainty and complexity. This is quite clear in the following quote from Funtowicz and Ravetz (1993): "Until now, with the dominance of applied science, the rationality of reductionist natural-scientific research has been taken as a model for the rationality of intellectual and social activity in general. However successful it has been in the past, the recognition of the policy issues of risk and the environment shows that this ideal of rationality is no longer universally appropriate" (754).

In simple words, if the accuracy of expert advice in a given issue is unknown or may be suspected to be low, it is not rational to place high confidence in that advice. The problem is, however, that our modern institutions do not yet distinguish well between the domains of policy making in which a predominant role for expert advice is appropriate, and those where it is not. From this perspective, the task is twofold: Develop new insights and institutions (for producing new kinds of knowledge, if possible; and for new kinds of policy making, typically referred to by terms as "governance" or "broad visions of governance"), and simultaneously emancipate the public from the old myths about the place of science and experts in society. It is a work of demythification of science, not of its denunciation, as noted already by Jean-François Lyotard ([1979] 1984). This I also believe to be quite clear in Nowotny, Scott, and Gibbons's text (2001). When they speak of the "epistemological core," this has no address in the scientific communities. I have never heard a practicing natural scientist use the term "epistemological core." In fact, I have barely heard them use words such as "epistemology" or "truth." Above, I quoted Funtowicz and Ravetz (1994) as having written that "in [postnormal] science, science is no longer imagined as delivering truth, and it receives a new organizing principle, that of quality." Four years earlier, however, the same authors provided a description of "normal" scientific practice that cut right through the truth rhetoric: "When we think of scientific knowledge as produced by craft skills which are shared and presupposed in an expert community, the categories 'true/false' and 'verification/falsification' reveal the severe limitations of their explanatory power. To be sure, good scientific work has a product, which should be intended by its makers to correspond to Nature as closely as possible, and also to be public knowledge. But the working judgements on the product are of its quality, and not of its logical truth" (Funtowicz and Ravetz 1990, 30).

There are differences between the practices of "ordinary" science and postnormal science, but perhaps the main contrast is that which runs between the actual diversity of practices, on the one hand, be they "normal" or "postnormal," and the myths and inappropriate images of science, on the other.

Philosophy as Progressive Politics

I have not yet been able to explain why the authors of the cited texts as well as many studies of contemporary philosophy; sociology of knowledge; science, technology, and society; and actor-network theory have felt the need to make fierce attacks on logical empiricist philosophy and its descendants. Why do Nowotny, Scott, and Gibbons (2001) attack the idea of an "epistemological core" if it plays no role in the discourses of scientists or policy makers? I cannot offer a fully satisfactory answer to this question. I have a hypothesis based on my personal point of view, but more debate is needed to clarify the issue.

It seems to be clear that the accusations against inappropriate images of science often come together with accusations against philosophy and philosophers of science. In part, this is to be seen as normal academic debate, in which one puts forward one's own position by attacking the positions of others: There is to some extent a basic disagreement about how to describe science in an appropriate and adequate way. Furthermore, I think that the rhetoric that includes our "sweeping claims" has to be seen as part of a polemic directed toward a policy-oriented audience, who may not have an explicit philosophical position but act as if they were convinced about the utmost superiority of all kinds of science. One may try to force them into debate by constructing a "positivist" straw man. In either case, the readers may still be puzzled by the energy put into the argument. First, there are not so many die-hard logical empiricists left, and it is not by any means obvious that they would disagree with, say, Funtowicz and Ravetz's argument about rational decisions under strict uncertainty and ignorance. Second, with few exceptions, the kind of philosophy under attack is not very visible (anymore) in the arenas of policy and governance. Rather, it has retreated to university departments, studying its own problems of how to reconstruct and conceptualize scientific theories.

Indeed, my hypothesis—again, merely based upon my personal experience—is that the effort put into the attacks on philosophy of science may be in part related to a sensation of intellectual disappointment. The reader may find my mention of an emotional issue inappropriate; however, I think that the experience of emotions should not be excluded from the discourse, especially in the context of discussion of science and values. This disappointment, I think, is related to an ideal of philosophy as a contributor to progressive politics or at least as an intellectual activity whose practitioners are committed to striving for a better world. Indeed, the large majority of our philosophical "heroes"—one could mention Socrates, Plato, Aristotle, Bacon, Descartes, Kant, or Mill—all supported social causes that were related to the content of their philosophy. This is undoubtedly also true in the case of the French positivists of the nineteenth century, as well as the logical positivists of the twentieth century. Regarding the latter, their focus on the singular aspect of rational reconstruction and the so-called "context of justification" was not at all arbitrary but played a role in an ideological battle against irrationalism, pseudoscience, superstition, and authoritarian politics.

Western history, however, is full of examples of philosophies and ideologies once recognized as progressive coming to be perceived as irrelevant or reactionary, as philosophy advances and society evolves. Nowadays, it appears from a certain perspective that part of the old "Enlightenment" philosophy, however adequate for its original purpose, stands in the way and takes on a reactionary role with respect to the political changes necessitated by the new problems of sustainability, because it does not offer a sufficiently nuanced image of science and its relationships to society. At the same time, many among the community of professional philosophers seem to be absent from the polis, devoting their efforts to theoretical problems originating in logical positivism/empiricism, perhaps not even anymore attached to the social cause that shaped them. Becoming even more personal and emotional, it startles me to see young philosophers in a conference on science and values focus their attention on "non-cognitive norms such as elegance in instances of theory choice," while from my perspective, "science and values" is about the role of science in the overexploitation of the Earth; the place and power of scientific expertise; the values at play, including economic values, in the setting of the research agenda, and so forth. I believe it was this kind of disappointment that made Gilles Deleuze and Félix Guattari (1994) accuse analytical philosophy of having become "infantile and perverse." For Deleuze and Guattari, philosophers ought to leave to the scientists the puzzles of making adequate descriptions, while the sole, but crucial, role of philosophy is to construct new concepts that may open new spaces for action, with the purpose of improving life and society.

The various strands and schools of scholarly studies of science in fact coexist and should be able to profit from their coexistence. Nothing would please me more if my emotional confession would provoke somebody to teach me about the causes, commitments, and stakes as seen from the "other" point of view. Obviously, the fight against irrationalism and superstition is not won once and for all.[3] It is also necessary to discuss the possible side effects and pitfalls of the discourses on socially robust knowledge and postnormal science in this respect. That, however, would probably lead us directly to an attack and not a defense of these concepts, a task I should leave to those who are better qualified for it.

NOTES

1. The concept of *Wissenschaftstheorie* is termed *vitenskapsteori* in Norwegian, *vetenskapsteori* in Swedish, and *videnskabsteori* in Danish. Indeed, for lack of a better translation, my own affiliation, *Senter for vitenskapsteori,* was given the official English name "Centre for the Study of the Sciences and the Humanities." At stake in the translation was the desire to indicate the inclusion of both natural science and social sciences, including humanities, in the concept of "vitenskap," and to include any relevant academic study in the concept of "teori."

2. See, for example, the Worldwide Virtual Network of Young Practitioners Working on Science and Society Issues, http://alba.jrc.it/ibss (accessed 10 July 2007).

3. This text was written in 2005, the fourth consecutive year in which this author's country, Norway, was run by a government led by Kjell Magne Bondevik, a Protestant clergyman. Others will have to speak for political developments in other countries.

REFERENCES

Cañellas-Boltà, Sílvia. 2004. "Uncertainty and Local Environmental Management—the Case of Maintenance Dredging of Polluted Harbours in Norway." Master's thesis, Universitat Autònoma de Barcelona.

Cañellas-Boltà, Sílvia, Roger Strand, and Barbro Killie. 2005. "Management of Environmental Uncertainty in Maintenance Dredging of Polluted Harbours in Norway," *Water, Science and Technology* 52 (6): 93–98.

Canguilhem, Georges. [1966] 1989. *The Normal and the Pathological.* New York: Zone Books.

Deleuze, Gilles, and Félix Guattari. 1994. *What Is Philosophy?* London: Verso.

Foucault, Michel. [1963] 2000. *The Birth of the Clinic.* London: Routledge.

Funtowicz, Silvio, and Jerome Ravetz. 1990. *Uncertainty and Quality in Science for Policy.* Dordrecht: Kluwer Academic Press.

————. 1993. "Science for the Post-normal Age." *Futures* 25:739–55.

————. 1994. "The Worth of a Songbird: Ecological Economics as a Post-Normal Science." *Ecological Economics* 10 (3): 197–207.

Gibbons, Michael. 1999. "Science's New Social Contract with Society." *Nature* 402: C81–C84.

Gibbons, Michael, et al. 1994. *The New Production of Knowledge: The Dynamics of Science and Research in Contemporary Societies.* London: Sage.

Hundertwasser, Friedensreich. 1997. *Hundertwasser Architecture: For a More Human Architecture in Harmony with Nature.* Köln: Taschen.

Husserl, Edmund. [1954] 1999. *The Crisis of European Sciences and Transcendental Phenomenology.* Evanston: Northwestern University Press.

Latour, Bruno. 1988. *The Pasteurization of France.* Cambridge, MA: Harvard University Press.

Lyotard, Jean-François. [1979] 1984. *The Postmodern Condition: A Report on Knowledge.* Manchester: Manchester University Press.

Nowotny, Helga. 1999. "The Need for Socially Robust Knowledge." *TA-Datenbank-Nachrichten* 8 (3/4): 12–16.

Nowotny, Helga, Peter Scott, and Michael Gibbons. 2001. *Re-thinking Science.* Cambridge: Polity.

Ravetz, Jerome. 1971. *Scientific Knowledge and Its Social Problems.* Oxford: Clarendon Press.

Strand, Roger. 1999. "Towards a Useful Philosophy of Biochemistry: Sketches and Examples." *Foundations of Chemistry* 1:271–94.

————. 2000. "Naivety in the Molecular Life Sciences." *Futures* 32:451–70.

————. 2001. "The Role of Risk Assessments in the Governance of Genetically Modified Organisms in Agriculture." *Journal of Hazardous Materials* 86:187–204.

————. 2002. "Complexity, Ideology and Governance." *Emergence* 4:164–83.

Strand, Roger, Ragnar Fjelland, and Torgeir Flatmark. 1996. "In Vivo Interpretation of in Vitro Effect Studies with a Detailed Analysis of the Method of in Vitro Transcription in Isolated Cell Nuclei." *Acta Biotheoretica* 44:1–21.1.

8

THIRD WAVE SCIENCE STUDIES
Toward a History and Philosophy of Expertise

CHRISTOPHER HAMLIN
University of Notre Dame

SOCIOLOGISTS HARRY COLLINS AND ROBERT EVANS call for a startling departure in science studies, the pursuit of "a normative theory of expertise" through a new form of inquiry, "Studies in Expertise and Experience" (SEE) (2002, 237). With regard to controversies ranging from the acceptability of human cloning to the safety of genetically modified foods, they urge a scholarship that will reconcile the competing demands for "extension," that is, a public say in policy, and for technical legitimacy. This would be done by distinguishing expertise and adjudicating its application. Ideally, the circle would be squared: controversies would evolve and resolve in a way that was appropriately participatory as well as fully responsive to the state of knowledge.

The call for an engaged science studies is not new (Woodhouse et al. 2002; Martin 1996; Richards and Ashmore 1996), but these authors go further than many in outlining an approach that is general and analytical rather than merely activist. They point to some of the determinations such a scholar must make—chiefly, that of demarcating expertise itself. Quite quickly the proposal drew forth a vehement response (Jasanoff 2003a; Rip 2003; Wynne 2003), to which Collins and Evans replied (2003). In various ways, the critics objected that any approach to validate the technical came at the cost of extension. In general, the SEE initiative has not been followed up. More recent scholarship has engaged with public policy mainly in terms of the institutions rather than the substance of expertise, and with the extension problem rather than that of technical legitimacy (Libertore and Funtowicz 2003). Indeed, it seems sometimes to be suggested that expertise is reducible to adequate mechanisms of public participation.

Yet expertise is more than institutions; hence the broad problem that Collins and Evans posed remains. My interest here is with how a broadened study of science, technology, and society, in which philosophers and historians (of technology, medicine, and the environment, as well as of science) were more prominent, might address that problem.[1] The sort of history and philosophy of expertise I suggest is a radical one, however. We cannot simply translate "expert" as "specialist scientist," or presume that a philosophy of expertise will be a simple derivation of the philosophy of science. What will be required is a reconceptualization of relations between science and its presumed areas of application— technology, medicine, and the environment—and a reorientation of the philosophy of science that will downplay some traditional issues, while highlighting others. Finally, it will involve a new view of the expert's professional obligation: generality, conclusiveness, skepticism, theory, objectivity, and disinterest may need to give way to the exploration of multiple scenarios, the provision of options to a client or in service of particular values, and to application in particular situations in which obtaining a high degree of certainty is not feasible, and in which it is recognized that any actions will benefit some and disadvantage others; expertise will be political.[2]

SEE differs most sharply from the sociology of scientific knowledge (the second wave) in treating expertise as real and discernible. By representing science as not fundamentally different from other endeavors, the sociology of scientific knowledge undercut (or seemed to) the ability to distinguish the expert. The new study, however, will treat expertise as an analyst's rather than an actor's category; that is, expertise will cease to be merely a claim made by a member of a cognitive elite, it will be a designation made objectively according to defensible criteria (Collins and Evans 2002, 240–41). No longer will the issue be whether or not there can be demarcation criteria; rather it will be where the boundaries properly lie. Ideally, scholars of expertise and experience will be able to provide a prescriptive delimitation of competence for the resolution of various species of controversy.[3]

A major thrust of Collins's and Evans's paper is that, in any controversy, the set of experts, those who really know, will not conform to the credentialed elite. Where the first (pre-Kuhnian) wave of science studies tended to treat all doctors of science (licensed bearers of the scientific method) as having more or less equal authority over all laypeople,[4] and the second wave assumed that no one had any more authority than anyone else, Collins and Evans argue that one must limit the appellation of expert to some small set of credentialed authorities (certainly not all doctoral scientists, even within a discipline), and, perhaps, to some subset of the uncredentialed as well, who are experts by experience (250–51). Others, both credentialed and uncredentialed, may have political bases for having their views considered, some strong, some more remote, but only these experts are to be accorded standing in the adjudication of the technical issues. Collins and Evans hope that the recognition that it is only a few laypeople who might qualify as experts by experience will make scientists less apprehensive about public participation, which some have hitherto viewed as the opening of the "floodgates of unreason" (237). Once they feel protected from attempts to swamp their expertise with democracy, scientists will avoid making simplistic and readily falsified declarations of authority.

It should be said that the argument made by Collins and Evans draws heavily on a single, iconic study, Brian Wynne's insightful reading of the "misunderstood misunderstandings" between government nuclear scientists and the sheep farmers of the Cumbrian hills following the 1986 Chernobyl nuclear accident (Wynne 1996).[5] In the Cumbrian episode, failure to combine two forms of expertise—that of the nuclear scientists, whose job it was to monitor Chernobyl fallout and protect the public, and that of the sheep farmers, who knew a good deal about how that fallout might (or might not) get into the sheep—resulted in a fiasco, which, Wynne suggests, unnecessarily threatened the farmers' livelihoods. On two questions in particular, the approach of the official experts, the nuclear scientists, proved inadequate. The first was their prediction that radioactive cesium would quickly bind to clay and thereby disappear from the food chain. This expectation rested on an inapplicable generalization; it applied to alkaline, not to acidic soils. The cesium hung around. That necessitated an indefinite restriction on lamb sales and a financial crisis for farmers, who had accepted the scientists' assurance that the problem would quickly resolve itself. Second was a study to determine the utility of the application of bentonite (clay) to render the cesium inert. Here the problem was that the test plots used in the trials bore no relation to the feeding patterns of sheep.

More broadly, the scientists' advice to the sheep farmers reflected ignorance about the economics and ecology of sheep farming, and even about what sheep ate. Wynne and Collins and Evans imply that this need not have been so. Recognition of particularly contaminated places combined with a fuller knowledge of the patterns of sheep grazing and, perhaps, more comprehensive monitoring of the sheep themselves might have allowed the industry to continue more or less normally. The expertise involved would have been more flexible, responsive to incoming knowledge from a variety of sources.

Accordingly, Collins and Evans argue that the sheep farmers should have been granted status, along with the nuclear scientists, as technical experts, not simply as stakeholders with political rights (2002, 249). They recognize why this didn't happen: not only was the farmers' expertise uncredentialed, it was not readily convertible into forms scientists will recognize. But it is here that the student of science and experience might have stepped in. By recognizing the necessity of drawing on disparate forms of "contributory expertise" (that of the scientists and that of the farmers), that scholar would be deploying "interactional expertise" in helping the farmers to secure a standing in the matter and to translate their experience into a common idiom (254–56, 261). Collins and Evans also identify another expertise, "referred expertise," that which is possessed by the successful director of a large laboratory who must assess contributory expertise in fields well beyond that in which he or she has contributory expertise (257).[6]

As well as expanding the designation of expert to include some of the uncredentialed (262), Collins and Evans also reduce it among the credentialed. For any question, it will be only a small subset of credentialed scientists who are properly accorded the status of expert. Their critique here is of an approach evident in much of the public understanding of science movement, which reifies "science" and draws an uncrossable gulf between the scientist's perspective (a consequence of a presumed shared background or ethos) and the layman's. Drawing on studies of the "esoteric sciences," those cases in the modern physical sciences which were taken as the exemplars of the social construction, they argue that for any question there will be a small core set of "scientists deeply involved in experimentation or theorization which is directly relevant to a scientific controversy or debate" (242). Thus, in gravity wave research, one's views will count only when well-recognized criteria of training and access to facilities have been met.

The result of this reconfiguration of how expertise should be recognized—

the distillation of science into a small core set, the inclusion of experience regardless of credentials, and the recognition that several distinct kinds of expertise must be deployed and coordinated—will allow a process to occur in which controversies will be resolved in a way that is both technically responsible and acceptable to the polity. Collins and Evans assert that this resolution is precisely what we should want; to deny "that scientists with experience of an esoteric specialism are the best people to make judgments about what should count as truth within that specialism" is simply incompatible with the values of Western civilization (243). These controversies will be of many sorts—in the assessment of new technologies and the planning of new installations, as well as in a variety of scientific controversies. The latter include the sorts of experimental controversies that Collins and Pinch have called "Golem" science; in historical scientific controversies in which the effects of long-term changes are being considered; and in the historical reflexive sciences, where those changes are the product of human action (266–68). That we could come to manage all these controversies in this way they acknowledge to be naive, but assert that academics have a "duty to be naive from time to time" (262–63).

The issues incompletely addressed in a finite paper on so large a subject are many. Is the approach suited to the full range of controversies? Can one truly sever the technical from the rest? Do experts settle controversies, or do we settle them in choosing experts? Should the Cumbrian sheep farmers be the archetypal case? Will we usually find, and need, experienced-based expertise, or were the government's experts in this case simply incompetent?[7] And, most importantly, *what is expertise?*

To a great degree the last question comprehends the others. For Collins and Evans, the most uncontroversial form of expertise was that of the specialized scientist, the member of the core set. That core set may well be the product of battles to bound a problem and delimit the methods used to solve it, and, as those battles may persist indefinitely, that core set may be more a conceptual than a social entity, but at least conceptually the equation of expertise with the core set seems plausible, perhaps indeed truistic. Collins and Evans defend that departure point with the "hard case" argument. In the second wave, one argued the importance of social factors by showing their influence where they would be least expected: in those esoteric branches of the physical sciences (2002, 242). If these were the hardest cases, it followed that others must be easier, that what prevailed there, in this case the emergence of a core set, must prevail elsewhere.

Yet what was the hard case for second wave science studies turns out to be a relatively easy case for the third wave, for here we are asked to work in the opposite direction: rather than finding the social in the rational, we are here attempting to locate something like "old-fashioned" epistemic competence in controversies that are far more mixed and messy even than Golem science. What were hard cases become oversimplifications. Thus the identity of any single core set is frequently precisely what is at issue in controversies that involve complicated and shifting criteria and context, numerous disciplines, incommensurable methodologies, and multiple subproblems that can go together in different ways. Where public good is at issue, the boundary problem is likely to be overwhelming and permanent.

In their response to such criticisms, Collins and Evans ceded the claim of core-set centrality. It was a convenient but inessential departure point for addressing the greater problem of identifying expertise according to the timetable of public need rather than the much slower one of scientific certainty (not to mention the even slower one of philosophical, sociological, or historical reconstruction) (Collins and Evans 2003, 442, 435–36; cf. Rip 2003, 419–21). And yet a great deal of the initial construal of SEE was bound up in that departure point. Jettisoning it transformed the argument in a way that Collins and Evans only began to recognize in their response: the expertise so desperately needed would be, they announced, "essentially imprecise (not-truth-like)" (Collins and Evans 2003, 435–36, 448).

The endorsement of a "non-truthlike" expertise may well baffle. Aren't knowing and knowing that you know the very essence of expertise? And what's to become of the centuries-old project of epistemology, the securing of knowledge and the central enterprise of the philosopher of science, if expertise is not like truth? Or is what is being suggested only some sneaky redefinition, some Polanyian subterfuge wherein knowledge is real enough but never fully extractable and inscribable?

While one might be able to characterize the several sorts of expertise that SEE will take up as forms of Polanyian knowing, to do so would miss the point, which is that the problems of knowledge and the problems of expertise are quite different. For what is proposed is a radical decentering of epistemology: at issue in many of the controversies is not whether knowledge is social or rational, but its very primacy. Doing may matter more than knowing, trust more than reason, prudence than proof, multiple options more than universals, shared risks more than individual insights, flexibility and adaptiveness more than comprehensive and authoritative master plans derived from first principles. In short, pillows and bolsters are to replace procrustean beds.

This decentering of epistemology brings with it the need to rethink the relation of the several kinds of controversies that require expertise. In presenting the core-set experts in the physical sciences as the proper departure point, Collins and Evans (in the first iteration of their study of expertise and experience) were employing a familiar perspective of the application of science. The assessments of technologies, the controversies over new construction projects, the disputes about whether humans were causing climate change and what should be done about it, were all presented as qualitatively like Golem science. The hard-case approach presumed that other, easier, cases are like the hard one, that in solving the one we have solved the others. This is the logic of applied science, that having solved the problem within the friendly confines of the laboratory and among our civil and rational scientific peers, we have really solved the problem. We venture into the messy world simply to enlighten. There we may well find more complexity, physical and social, and even wholesale irrationality and various forms of know-nothingism, but the problem remains solved; always the onus is on the context to yield to the rigor of the laboratory. In such a view, matters technical, medical, and environmental will be seen as properly derivative of science. Except on the margins, or as afterthought or as excuse for screw-up, there is no intrinsic place here for lay expertise.

But if expertise is to be no longer truthlike, this pecking order is indefensible. To wait for certainty is to wait for Godot. Everyone else has gone on changing the world without it (though not without assuring the gullible that they have it). One resolution to this situation would be to demarcate ever more sharply: perhaps the slow routes by which expertise overcomes disagreement in science really have nothing to do with the messy normative issues of policy.[8] But another, which I follow here, is to ask whether we have gotten the right Ur-form of controversy—that from which others are seen as departures manifesting certain characteristics more or less conspicuously. Some years ago I suggested that in mixed and messy controversies (those, broadly speaking, that include science as well as technology, health, or environmental matters; that involve normative and moral questions, and require making decisions in the absence of perfect knowledge), the perspectives of technology studies offered the better departure point.[9] Rather than representing technology as derivative of science—the usual applied science approach—it seemed better to start with the mix and the mess: the complicated relations and tensions between knowing and acting appropriately called technoscience (Jasanoff 2003b, 159–61, 2003c; Wynne 2002). Recent recognition of the prominence of Mode 2 knowledge production or "postacademic" science is consistent with this approach. "Science," it is being recognized, is more than ever about doing rather than knowing (and much of that doing is about money making) (Gibbons et al. 1994; Ziman 1996). This is not the place to review the general features of the argument that it makes more sense to view science as a peculiar variety of technology rather than the reverse, but only its implications for expertise and controversy.

One place where that perspective does come to bear is in the common view, shared by Collins and Evans, of controversy as a pathological state to be cured by expertise. They assume that science will resolve contested questions and reduce disagreement, though premature intervention into the work of the core set either by the broader scientific community or by the public at large may truncate the convergence the experts are working toward and impose either a different resolution or politicize matters into a state of permanent controversy. From the viewpoint of science, controversy represents the problem not yet solved, the consensus not yet arrived at, the textbook that cannot be written and, perhaps, the expert advice that cannot rightly be given. They share this view with philosophers of science such as Kitcher. It seems to involve the following propositions:

1. Controversy is essentially illegitimate and ideally transitory, since there is a right answer.

2. Controversy will be resolvable by recourse to a nature that exists prior to and independently of the controversy.

3. The problem will be solved when the right people have enough time and sufficient resources.

4. Science, if unavoidably political, can and should be effectively independent of the several positions in the controversy.

5. The inertial state of controversy is toward convergence or even closure, a term which connotes a civil and satisfactory wrapping up.[10]

In fact, however, this is a difficult view for the holder of a science-centered approach to maintain consistently. For it seems to ignore the fact that science is often the source of controversy, as well as being its fuel and its product. Many of the controversies Collins and Evans want to consider—for example, ozone holes or global warming—would not exist had not scientists triggered them.

In mid-nineteenth-century London it was the new availability of means of water analysis that generated a controversy about water quality—reform efforts three decades earlier had fizzled because there was no plausible persuasive language for a critique (Hamlin 1990). And at the same time as they have deplored controversy, philosophers of science have worshipped dissent, insisted that theories and research programs be assessed in terms of their fertility to generate novelty. Michael Polanyi held that the good scientist would engage passionately in controversy and would know when to hold a conjecture in the face of evidence against it. James Conant famously declared that the scientist's commitment was ever to destabilize what we thought we knew; one's goal was not knowledge but discovery: the day that all was in the encyclopedia was the day science would die.[11]

Controversy has also been welcomed by some of those concerned with public participation and accountability (Wynne 2002). They see it as a sign of democracy resisting dogma. Conant, writing more as research administrator than historian-philosopher, even came up with a scheme for manufacturing dissent if it did not offer itself (Conant 1952, 67–69; see also Mazur 1973). But the celebration of controversy is no less troublesome than its repudiation. Controversy can as easily be the manifestation of dogma resisting democracy—assuming, as will rarely be the case, that the positions of democracy and dogma are clearly distinguishable and readily equate to those of good and evil. If the stakes are high, threatened parties may invest in countervailing expertise. While the pretense invariably is that the science is being carried out to settle the questions, it is often more correct to see it as "adversary science," the endless feeding of the insatiable demon of epistemic exhaustiveness. So long as a question can be kept in the epistemological ballpark, unpleasant policy can be postponed. Thus, a Bush administration advisor counsels the president on how to discuss global warming: "Should the public come to believe that the scientific issues are settled, their views about global warming will change accordingly. You need to continue to make the lack of scientific certainty a primary issue in the debate" (Burkeman 2003; see also De Marchi 2003).

As David Collingridge and Colin Reeve observe, how much science will be brought forth and how long it will continue depends on the resources committed to its creation rather than on the nature of the question at hand. They conclude that "no choices of policy are ever made which are sensitive to any scientific conjecture, and that no such choice ought to be sensitive to any scientific hypothesis" (Collingridge and Reeve 1986, 28; see also Nowotny 2003). Allan

Mazur goes further to recognize that "a technical controversy sometimes creates confusion rather than clarity and it is possible that the dispute itself may become so diverse and widespread that that scientific advice becomes more of a cost than a benefit to the policy maker and to society." His solution is to find "reasonable procedures"—formal means of controversy resolution—but this seems incompatible with the freedom of the inquirer to continue to dissent, even at the cost of marginalization.[12]

The science that emerges in these circumstances of orchestrated controversy is not necessarily bad science. It is likely to be uncommonly exhaustive and may well be innovative; continual disagreement requires a nurturing social context no less than orthodoxy, notes Nowotny: "Controversy may simply be the forcing bed for a Popperian paradise of conjectures and refutation" (Nowotny 1975, 35–36; Hamlin 1986). In the London water controversies of the late nineteenth century, those under the most pressure for the pollution of the Thames and for supplying contaminated water to the public sought to win the controversy by out-data-ing their opponents—indeed, gathering data far faster than it could be adequately analyzed. But some of their work led to the first modern study of reoyxgenation dynamics of a large body of water, a very costly research project, which it is clear would not have been undertaken outside this context. It also led to a paradigm change in sewage purification (Hamlin 1988).

Often participants will be making the science as they go along, intentionally or accidentally, rather than downloading it from some Platonic heaven of pure knowledge. As they choose what problems to work on, how hard to work on them, by what means and toward what ends, they will be like the designing engineer or the craftsman, creating within a certain framework or template. The result is the science Robert Proctor describes: a great deal of knowledge of how to do some things (like finding oil) surrounded by a great deal of ignorance of other things, all masquerading as the product of the choices of value-free inquirers following up the problems of greatest importance (Proctor 1991; Jasanoff 2003c, 2003b, 159).

None of this fits easily with idealized views of science. In science this ambivalence to controversy arises because of its simultaneous fixation with universal truths and with endless progress, its privileging both of sure knowledge and of constant doubt. Problems are tackled according to their ability to provide the most general knowledge from which can be derived solutions to all the pesky problems that plague society. Technology is a much better place to look for a "non-truth-like" entity. It has no equivalent to epistemology, and no one waits for the arrival of general knowledge. Rather, technics adapts as expertise responds to ever changing preferences, opportunities, needs, and circumstances. Controversy—in the sense of a complex discussion among alternatives—is the normal state of affairs, whether or not one holds it to be the desired one. In short, we have no business presuming a tendency toward resolution through the application of expertise, since it may equally be the source and means of controversy. This is not to say that controversy need be permanent nor that some controversies cannot be closed rationally. Some parties may be squashed under the tank tracks of reason or exhaust their resources of resistance or simply tire of the issues (Engelhardt and Caplan 1987). Or, recognizing that partial agreement will be the best platform for effective dissent, partisans may move the conflict to another front (as happened in water analysis) (Hamlin 1990, 200–201). The conclusion one would take away from a technology-centered approach, then, is that what should be at issue is not controversy per se, but particular controversies—the options available for resolving particular problems.

Such an approach allows us to respond to one of the concerns of Collins and Evans—expertise of some form is available for deployment before the great questions are settled. (Indeed, one might go so far as to say that the deployment of experts is constitutive of controversies—Mazur [1973, 10] defines "'experts' as two or more people who can authoritatively disagree with one another.") It does not, however, suggest that the involvement of experts need tend toward convergence, nor does it meet their desire to recognize a transpolitical expertise: expertise, they assert, "must be treated as a category *sui generis* that is separable from its politics and its attribution" (2003, 448). To see how those claims fare, we need to look more closely at the combined issues of what expertise is and who the experts are.

To adjudicate and allocate expertise requires recognizing it. Collins and Evans confine contributory expertise to the core set, supplemented by laypeople with appropriate experience. Among critics, Arie Rip (2003) in particular found this unsatisfactory. It missed the seriousness of boundary problems, failed to provide a broadly applicable concept of expertise.[13] Indeed, the treatment can tend toward tautology. In the gravity wave controversy to which Collins and Evans allude, it will be the case that core set and controversy mutually define one another: that is, the gravity wave controversy is the set of issues that gravity wave researchers worry about; the gravity wave core set are those engaged in disputing about gravity waves.[14] There may be laypeople or other scientists with an interest in the subject, but there are no obvious external stakes for them (at least not now—three hundred years earlier the ontological basis of gravity did have important metaphysical implications) (Kitcher 2001, 67–68). It is not so much that one should expect an expert to be good at doing something other than his or her narrow specialty, more that this equation offers no basis for adjudicating and allocating, no leverage for understanding why the core set is composed as it is. Identifying it with unique instrumentation and training begs the question of why these should have been privileged. Here we are stuck with expertise as an actor's category, precisely what Collins and Evans had hoped to avoid.

While one need not deny that there are core sets even if one cannot always justify them, many controversies will be between rival core sets, or will reflect core-set imperialism. Where the controversy itself is a struggle to define expertise, the "leave it to the core set" approach will beg the question. A familiar example is the controversy over DDT. Who were the experts? Entomologists, ornithologists, epidemiologists, or even some new form of environmental chemist that was yet only a twinkle in the eye of the god who designates disciplines (Graham 1970).

Even the seemingly straightforward matter of disease causation, which would seem to demand medical science, invites multiple approaches, which may bring with them quite different kinds of resolution. In the cancer wars, the struggle has sometimes been between core sets of epidemiologists and of pathologists—between the methodological disciples of John Snow and of Claude Bernard. But even within epidemiology there are distinct core sets. John Snow's 1854 demonstration that water from the Broad Street pump transmitted cholera in Soho is a familiar example of inductive inference as well as applied expertise, but, as I have argued elsewhere, Snow's achievement privileged some kinds of questions, assumptions, methods of proof, and interventions over others. Snow, a single-factor epidemiologist, was interested in what precipitates a case of disease and by what means a disease moves. John Sutherland, who, along with most sanitarians, was interested in the interaction of multiple factors, was interested in what initiates an epidemic and what causes X rather than Y to sicken when both are exposed to pathological forces.

Both Snow and Sutherland were engaged in scientific research programs, both recognized water as an important factor, but bound up in the privileging of some questions were assumptions about how institutions contributed to the creation of disease and at what sites medical intervention was most effective and appropriate. What in society was fixed? What would give?[15] Moreover, neither approach implied a single solution. Snow's findings, which have been taken as the scientific foundation for public provision of safe water, were equally compatible with extreme individualism: that each person would take direct responsibility for the safety of the water he or she imbibed. The traditional Chinese solution to water quality problems was that all drinking water be boiled (Hamlin 2001). Nor was this a transitory condition. Modern epidemiology includes those who concentrate on narrow solutions to specific problems of parasites and infectious diseases, and those like Richard Wilkinson, who work with the broadest determinants; Wilkinson shows mortality differentials to be a function of inequality (Kawachi, Kennedy, and Wilkinson 1999). These groups have quite different senses of the public health agenda, though neither is monolithic. What this should suggest is a picture of a complex dynamic network of institutions, multiple applicable core sets to particular problems, and the generation of multiple options even by members of a single core set. One may sometimes narrow options in choosing the form of expertise to deploy, but that choice rarely determines a single option.

Moreover, one cannot expect the various groups of experts to be sitting cozily on the shelf, each tagged with a sign designating its appropriate domain of application, methodological loyalties, and assumptions about social action, so that we poor members of the public can readily find the one we like. On the contrary, some sociologists, like Andrew Abbott and Ulrich Beck, depict a landscape populated by rival groups eternally struggling against one another for authority over some bit of cognitive domain (Abbott 1988; Beck 1992). Perhaps the most brazen examples of core-set imperialism were the classical political economists of the first quarter of the nineteenth century. Remarkably, the group of analysts associated with the *Edinburgh Review*, later with the *Westminster Review*, and with the Political Economy Club had neither formal training nor a professional body to credential them (though through Adam Smith, Lord Kames, and Dugald Stewart they had tenuous links to Scottish moral philosophy). And yet, as Maxine Berg (1980) has shown, very rapidly they became the uncontested authorities in a wide range of public policy matters. Other candidates would be the expansion by physicists into foreign policy during the cold war and of Freudian psychoanalysis into social work (York 1987; Herken 1987; Lubove 1971). None of this should suggest that the motives of these thrusting core sets were venal (though that question certainly becomes more relevant as commodification pervades the academy); political economists, arms physicists, and psychoanalysts had good reasons to think that their participation in public life was essential for prosperity, survival, and human happiness.

This picture of a complex and open-ended application of expertise is in sharp contrast with the common image of the expert as hyperspecialized, the person who knows most about least. In light of that recognition, how should we reconceive expertise? What should the expert offer? While the terms "expert" and "expertise" are used even more loosely than are "scientist" and "scientific knowledge," they do suggest some important distinctions. Unlike mere "knowledge," which can be general, "expertise," like "application" implies ends. We may say that X is knowledgeable, but that Y is an expert heart surgeon, negotiator of contracts, tightrope walker, or computer designer—skills whose desired outcome is plain. If expertise can only be measured in terms of ends, it means that we have to consider what ends are at stake in controversies. Is there something that is trying to happen in the controversy? Something that should happen? Do those involved agree about ends? Mazur's application of the term "expert" to those who authoritatively disagree is simply a subset of this expertise-as-action for cases in which the parties do not agree—perhaps about ends, perhaps about means.[16]

The conceptualization of expertise in terms of ends has implications for the rhetoric of expertise. It has often been suggested that the expert's proper domain was to inform, leaving it to the appropriate public bodies to determine how best to act on that information. This perspective was central to the so-called "deficit model," that expertise was the speaking of truth to power (De Marchi 2003, 175; Nowotny 2003, 165). But this ethic of expertise is not and cannot be enforced. Even in the best of circumstances it is impossible to fix the spot where description leaves off and persuasion begins. Mazur writes: "A technologist or scientist soon comes to recognize that the complex technical problems of the state of the art require subtle perceptions of the sort which cannot be easily articulated in explicit form. When it is necessary to make a simplifying assumption, and many are reasonable, which simplifying assumption should be made? When data are lacking on a question, how far may one reasonably extrapolate from data of other sources? How trustworthy is a set of empirical observations?" (Mazur 1973, 20–23; see also Jasanoff 1987, 200). This notion of expert as information provider is not only naïve; it is also generally inaccurate. That is, if we simply want to know facts, we consult a reference book. What we usually want from an expert, by contrast, is an assessment or an authorized solution that permits action. Need expertise as persuasion subvert meaningful public participation? I think not, but I return to that key question in the next section. If expertise is about achieving ends, we cannot expect experts to be passive; their expertise will be measured by their success. Let me offer a couple of examples from my own historical work. The first is the career of "x-pert" Edward Frankland (each member of the elite X-Club, which ruled mid-Victorian London science, had a nickname with the prefix *x*; Frankland's was X-pert). Frankland, between 1867 and his death in 1898, was the most important British expert on drinking water quality. He probably did more than any single person to secure safer water supplies during that period. I take this as Frankland's end; it was also the only publicly acceptable end in water controversies, though many of Frankland's opponents wanted to persuade the public that the water already was safe. In 1868 Frankland became the world's most accomplished water analyst by inventing a process that allowed him to measure organic nitrogen directly at an order of magnitude beyond any previous laboratory. Using this achievement to leverage an appointment as one of three (and soon two) Royal Commissioners on River Pollution, he quickly became the principal authority: by 1875 his laboratory had analyzed far more water samples than had any other.

Frankland recognized quite clearly that to measure organic nitrogen was not to measure whatever it was in water that caused fecal-oral diseases. He also realized two other things. One was that he knew better than almost anyone else the danger that contaminated water posed, the other was that in his laboratory he usually knew less about the quality of a particular river water sample than did a layperson who lived on that river's banks and saw what was dumped into it on a daily basis. Nonetheless he acquired enormous authority by expressing regularly, in technical, graphical, and metaphorical modes, what he regarded as common sense: that water which had been contaminated by human wastes should not be assumed safe to drink. He knew with a great deal of confidence that what was wanted was in the current state of science unknowable, but he did not eschew the obligation of the expert to provide authority. Nor was Frankland's intervention simple skepticism, much less a destructive display of ignorance, though his opponents (many of them in the employ of companies that distributed contaminated river water) were united in condemning him for violating the expert's presumed credo of simply stating the facts. In reaching conclusions on general principles rather than on the results of (irrelevant) tests that only a skilled chemist could perform, Frankland was being disingenuous and even dishonest, they insisted. Nonetheless, significantly on the basis of his authority, towns spent millions of pounds acquiring better water supplies, probably saving thousands of lives (Hamlin 1990, chaps. 6 and 7).[17]

The second case is of mid-Victorian civil engineering. A review of the history of the great projects of that period—railroads, reservoirs, aqueducts—will reveal some projects that failed utterly, and many others, indeed most, which came in over budget, often by several hundred percent, and late. It is striking also that the few engineers with better records, like Joseph Locke, were not at the highest tier of the profession. The reason is that those who were most candid about uncertainties might never get permission to build: another fast-talking miracle worker would have had the contract. If we measure expertise in terms of common scientific canons such as accurate prediction (or even honesty) the engineers will be found wanting. And yet the expertise of these engineers and of the financiers associated with them, men like George Hudson, was sufficient much of the time to get the works built. The successful engineer knew about iron and brick, soil and rock, but also about gullibility, flattery, and showmanship, and about the interests of other people—assistants, laborers, policy makers, share buyers, financiers. To be sure, one needed to find creative solutions to unanticipated structural problems, but also to mollify anxious shareholders. The great desideratum, as Anthony Trollope's odious Melmotte, the financier in *The Way We Live Now*, explains, is the ability to engineer confidence. Over the centuries, many doctors have argued that this applies in medical care as well.[18]

Leaving aside the question of whether these episodes represent acceptable forms of public participation, one may ask "are these people experts?" Can a scientist whose expertise reduces to the statement that "I know nothing more than you do" be an expert? Or an engineer whose real skill is inspiring confidence at the expense of the common virtue of truth telling? According to traditional canons of scientific conduct, the answer is "no"; with regard to effecting broadly progressive ends, it is "yes" (Nowotny 2003, 152). Both examples also reflect the illusoriness of severing the technical from the political (see Latour 2004). Human behavior is both part of the context of the technical problem and often part of the problem itself. The expert's new water pump will not work without a system to ensure its acceptance and maintenance, which may, incidentally, involve reengineering gender relations and power structure of a traditional village (Drangert 1993).

It is possible to conceive of a form of Science and Technology Studies (STS) that would act in much the way Collins and Evans envision, as complement, mediator, and occasionally critic of such forms of expertise (Wolfe et al. 2002). It would be part of a "civic epistemology" (Jasanoff 2003c, 2005a; Collins and Evans 2003, 437). Its components might include the following goals:

- analysis and evaluation of goals or at least outcomes acceptable to the various parties, and a decision about how to engage in the controversy—is one partisan, mediator, or even spectator?

- the ensuring of a full laying out of options, of the assumptions underlying estimations of the practicability of those options, of their compatibility with one another, and of the needs for knowledge and action associated with those options. How many ways are there to accomplish the goal? What forms of knowledge do they incorporate? What risks do they entail? (Beck's term, "specialized context research," suggests the image of the shunting yard, with its multiple tracks and switches, over the single train; Rip finds a precedent in the wide-ranging knowledge of the natural historian; Beck 1992; Rip 2003.)

- a consideration of the varieties of core sets which might be applicable; exploration of the implications of various forms of inquiry and of claims of certainty or uncertainty for public action.

- an ability to accommodate the full complexity of the controversy as the participants understand it rather than an idealized version that may be more susceptible to models. What institutions and actions do various expert interventions imply?

- ensuring a flexibility of decisions to accommodate changing circumstances, new knowledge, and new options (Beck 2002, 174).

Some of these attributes are more fully developed in European policy, where the precautionary principle holds considerable sway, than in American, but even European policy often seems more a piecemeal movement *away* from modes of expertise that have proven unsatisfactory, than a clear movement *toward* a particular conception of expertise and role of an expert. In some respects, a philosophy of expertise as prudence would look much more like thorough investment counseling or good medical care than natural science—one would need to be thinking simultaneously about goals, risks, and, as broadly as possible, about how to assess the determinants and impacts of future events and to influence those events (Jasanoff 2003c).

The investment counseling scenario or that of the construction of a regimen to secure health both recognize that the actions of experts are not merely descriptive and predictive. Experts—both scholars of science, technology, and society and others—are part of the system we are studying; our own actions contribute to the future we desire (Toulmin 1982). The focus on prudence would also give central attention to the expert's relation to the community. That is, like Frankland and the engineers, the expert concerned with achieving a particular end confronts a question that does not usually preoccupy the scientist: how to present oneself to others. The demeanor so important to the expert witness is no secondary criterion where trust is primary, where relationship matters at least as much as knowledge. Notably, prudence presumes no easily reifiable entity, like science's nature. Prudential judgments are conditional, and subject to continual modification. Nor is prudence usually embodied in any institution—we do not award an M.Prud. degree. Quite the contrary, rather than highlighting those factors in which experts differ from laypeople, the focus on prudence highlights determinations which must be made well or poorly by any individual or community. A philosophy of expertise so configured seems different from what ethicists or philosophers of science, technology, or medicine now usually do.[19] It would certainly draw in epistemology, ethics, and justice, but stretch beyond philosophy as usually conceived into rhetoric, on one end, and development studies on the other.

To see how such an enterprise might look in action, it will help to return to the Cumbrian sheep. In three respects, the expertise-as-prudence approach suggests a basis for integrating the knowledge of the nuclear scientists with that of the sheep farmers—precisely what Collins and Evans seek to do.

First it alerts us to the importance of goals. Are the goals of the parties identical or even compatible? Wynne and Collins and Evans presume that they are: all are interested in finding a way for the farmers to continue raising sheep in the wake of Chernobyl. And yet, the farmers are likely to be more concerned with the survival of their farms, while the scientists (notably from the Ministry of Farming and Fisheries) may be more concerned with the reputation of British lamb, the capacity of the British government to ensure the safety of exports, and even with the need for the government to be seen as decisive, than they are with the welfare of a few farmers. To them, careful pasturing and extra monitoring may be less attractive than a blanket ban. There may or may not be sufficient overlap in these goals to accommodate both, but one can hardly explore the accommodation until one has acknowledged the tension. My point simply is that it will make no sense to assess the expertise without assessing what is at stake for the parties involved.[20]

Second, we have, in the integrating concept of prudence, a way of recognizing that the expertise of the scientists and of the farmers is not in fact different in kind. The problem with the scientists' approach is not that it privileges their so-called expertise over the layperson's experience, but that it is bad science. The scientist who claims to have solved the problem of the oral uptake of cesium by sheep while ignoring how sheep eat or the institutions of sheep farming has solved no problem. Solving that problem does not necessarily demand extrascientific elements (though some, like Frederick Suppe and Wendell Berry, hold that the unique problems encountered in most farming operations will expose the inadequacy of all general knowledge; Suppe 1987; Hamlin and Shepard 1993); the farmers have just alerted the scientists to variables that need to be considered in a comprehensive expertise. But, more broadly, what the farmers brought—or equally what James Randi brought to the Benveniste laboratory—was skepticism, an attribute hardly incompatible with science (Collins and Evans 2002, 264–65; Picart 1994).

Third, and related, is the issue of the conformity of the experts' forms of assessment with the layperson's. Wynne notes that the farmers' reaction to the nuclear scientists is a matter of their perception of knowers and of the hallmarks of knowledge—that these are inextricable from the substance of the knowledge. In contrast with Collins and Evans, he highlights the ongoing problem of trust that the nuclear scientists face. This is more than a matter of their credentials, competence, or even their possible bias as ministry employees, though these will all contribute. In the Cumbrian controversy the most pronounced manifestation of differing ideas about knowledge is the response to uncertainty. For the scientists, to be publicly scientific is to be definitive. Doubts are only for the ears of other scientists. Unfortunately, the definitive pronouncements are wrong, and those who act on them are harmed. The farmers, however, don't expect or even respect certainty, and have a more complicated way of determining credibility.[21] To them, to lack doubt is a sign of naïveté or simple stupidity.[22] Hence it is not simply that the institutional form of the expertise is intrinsic to its assessment, but that cognitive components—emphases, for example, on simulations or on control and prediction; or that involve objectification of what cannot be objectified, like willful humans—may seem alien. The trust that farmers do grant is unlikely to be in any science per se, or in credentials, but rather in persons, and it will be partial and fragile. A long track record may count for a great deal more than a brief virtuoso performance (Wynne 2001; Ottinger forthcoming). Yet its components are hardly idiosyncratic or resistant to analysis.

Sadly, it is sometimes suggested that to take knowledge as seriously social is to treat it as venal, vulgar, and arbitrary, as if "cognitive" and "social" were antithetical. The move from the second to the third wave is predicated on this not being the case, and Wynne's scientists and sheep farmers help us understand how it is not. I have suggested what philosophers of science might do in all this, but what of historians? Mainly, I think the historian must be the resister of reification, by insisting on the fragility of confidence and inescapability of context.[23] Historians do this by focusing, as far as possible, on the unfolding present with the constant confrontation of problems to be coped with, not on presumably eternal nature. Without denying that certain components of scientific knowledge and procedure become black-boxed, their origins and warrants forgotten and inaccessible, the historian can nevertheless point to the assent in scientific authority that is continually necessary. Good historians, in short, take up the farmers' comprehensive integrating viewpoints: they know that expertise was (sometimes) real enough, but also that the applied knowledge is in (and not prior to) the application.

I have suggested that expertise is not simply an actor's category, but also that it cannot be reified and abstracted from situations—we cannot, as Collins and Evans suggest, simply add 3 ml of contributory, 2 of interactional, and 1 of refereed expertise; instead, expertise is real, embodied, and dependent on the changing levels of trust and credibility. While it is not clear to me that we have a philosophy of science of trust, Wynne suggests that trust and credibility will be based on much more than demeanor or haircut, that they are the outcome of an interaction between nature, the expert, and the expert's constituents. Components of expertise will include not only narrow technical competence but the ability to recognize the need for action, to recognize and criticize multiple options, to understand the political and social structures in which knowledge is to be applied, and to privilege practicability and flexibility. None of these are beyond analysis and assessment. Their study and application will constitute the third wave of science studies.

NOTES

1. The volume edited by Selinger and Crease (2006) appears to be an excellent contribution but arrived too late to be included in this analysis. Its contents include some of the articles I review here.

2. The approach is broadly in keeping with that of Proctor (1991) and Sheila Jasanoff (2003b, 2003c, 2005b).

3. This has been a theme of Collins's and Evans's more recent work in the Expertise Project at the University of Cardiff. See http://www.cardiff.ac.uk/schoolsanddivisions/academicschools/socsi/staff/acad/collins/expertise/index.html (accessed 23 November 2006).

4. A remarkable characteristic of the public face of science in the 1950s, at least in America and Britain, was its unity. See for example Barber (1952).

5. Wynne has raised some of these points in his response to Collins and Evans (Wynne 2003).

6. Collins and Evans believe that the science studies scholar has also some claim to a status as the coordinator of these various kinds of expertise: "The appropriate balance of contributory expertise, interactional expertise, and referred expertise has still to be worked out, and so has the work of discrimination and translation" (2002, 260).

7. In agriculture, something like experience-based expertise is relatively well recognized. See Hamlin and Shepard (1993).

8. I am struck that many volumes of collected papers tend to lump together controversies with some significant technical component. Gravity waves are treated in the same way as the viability or safety of new technologies. This is not only the approach taken by Collins and Evans, but that which characterizes the excellent 1987 Engelhardt and Caplan volume on *Scientific Controversies*. See also Machamer, Pera, and Baltas (2000). Jasanoff (1987, 202–3) notes by contrast that Alvin Weinberg's concept of transcience was an effort at separation. By outlining the many questions that science could not settle, however much it might seem to, Weinberg's paper provided a clear demarcation that liberated science from the contamination of policy.

9. Science-technology relations are treated quite differently by writers on science and on technology. Those who come from the side of science (for example, Kitcher 2001, 86–88) tend to presume the applied science model; those who come from the side of technology often employ the technoscience model, or Layton's "mirror image" twins. At the founding of SHOT, there was a semipublic rejection of the notion of technology as derivative, and the question of science-technology relations has hardly been at the forefront of the philosophy or history of technology (Hamlin 1992; Seeley 1995; Sinclair 1995).

10. The classic statement of this position is implied in the Bernard Barber's (1961) title: "Resistance by Scientists to Scientific Discovery." Brante (1993) sees this perspective as in part based on a particular moment in sociology, associated with Merton, Parsons, and Sarton, which presumed a tendency toward harmony and efficiency.

11. See Polanyi (1957a, esp. 115); Conant (1951, 6–13); Polanyi (1957b, 484) likened post-controversy science to the preparation of telephone directories. See also Nowotny (1975).

12. See Mazur (1973, 30, 40–41). Such attempts to create consensus or extract the

science from the policy have been objected to as insidiously favoring one side. See for example Markle and Chubin (1987, esp. 2): "the CD [Consensus Development] program acts as a mechanism to contain rather than to resolve controversy, and therefore, functions as a mechanism of social control: to confer upon scientific uncertainty the imprimatur of official agreement, authority, and policy. The conferences, rather than being sites of negotiation and conflict resolution, are formal and predictable." And later: "We conclude with some observations on the public performance of democracy, which has all the trappings of theater but the unmistakable mark of backstage politics." For more subtle aspects of these problems see Jasanoff (1987, esp. 218–20); Rayner (2003).

13. Others too have criticized the self-servingness of problem framing which is done only by the experts. See Jasanoff (2005b); Rayner (2003); De Marchi (2003).

14. Collins and Pinch (1993, chi. 5). The very term "esoteric" used by Collins and Evans (242) implies that the issues are defined by a core set of adepts and are intrinsically unavailable to a wider constituency.

15. See Hamlin (1990, chap. 5; 2002, 915–19). The selection of methods and the assumptions about society that go with them is a major theme of Beck (1992, esp. 174). "The prevailing theoretical self-concept of science implies that the sciences cannot make value judgments with the authority of their rationality. They deliver so-called 'neutral' figures, information, or explanations which are to serve as the 'unbiased' basis for decisions on the broadest variety of interests. *Which* interests they select, however, on *whom* or *what* they project the causes, *how* they interpret the problem of society, *what* sort of potential solutions they bring into view—these are anything but neutral decisions." Beck generalizes that where such decisions are being made "scientific cognitive practice becomes an *implicit, objectivized manipulation of latently political variables, hidden behind the pretense of elective decisions not subject to justification*" (170).

16. The following discussion draws much from Jasanoff 2003c, and 2003b, and 2005b and Turner 2001. However, I am more concerned to highlight the independence and the skills these experts bring.

17. Collins and Evans seem to find room for experts like Frankland: "One of the paradoxes of expertise is that it might be right to consult it even when the likelihood of it producing correct knowledge is almost zero" (2003, 448).

18. For engineers see Lambert 1934, Binnie 1981, Conder 1983. For this argument in medical practice see Harley 1999, 407–35.

19. Some recent commentators complain of the imperiousness of "moral expertise" (Nowotny 2003, 154; see also Wynne 2002). The study of inductive risk would loom large in such an endeavor. See Richard Rudner 1953; Douglas 2000.

20. A useful analysis of the differences between the goals of public experts and of the clients on whose behalf they ostensibly work is Edelstein 1988. Beck (1992, 173) and Collingridge and Reeve (1986), however, warn us not to expect much commonality of goals—the controversy that exists has frequently arisen over goals or interests.

21. An example might be the question—that was asked to Victorian engineers—of whether they owned shares in the project. In the 1840s engineers were expected not only to own shares but to meet calls for additional capital, which ruined some of them (Vignoles 1982). Similarly, professors of agriculture in nineteenth-century American "cow colleges"

were expected until late in the nineteenth century to be farmers themselves at the expense of developing specialized knowledge: how else could they pretend to be able to say anything about how to farm? (Marcus 1985).

22. Some recent commentators (Wynne 2002; Jasanoff 2003c, 237; Rayner 2003, 169) have taken issue with the supposed virtue of transparency. As the cases of Frankland and the civil engineers suggest, one need not privilege candor. One can make the case that the successful expert is not necessarily the most honest in representing doubt. Likewise, one need not assume that expertise is about presenting all options to a democratic polity. There may be situations in which the expert will decide to proceed by obfuscation in the support of oligarchy. The argument here is simply that the expert will not do well by representing knowledge in forms that are incompatible with the forms in which the recipient group conceives knowledge.

23. The framework in which experts' statements are received is central to Wynne's criticism of the propositional focus of Collins and Evans's concept of expertise. As Wynne notes, the UK controversy about genetically modified food is about much more than DNA; accurate statements of uncertainty will not change that. See Wynne 2002, 2003. For broader issues see also Jasanoff 2000.

REFERENCES

Abbott, Andrew. 1988. *The System of Professions: An Essay on the Division of Expert Labor.* Chicago: University of Chicago Press.

Barber, Bernard. 1952. *Science and the Social Order.* Chicago: Free Press.

————. 1961. "Resistance by Scientists to Scientific Discovery." *Science* 134:596–602.

Beck, Ulrich. 1992. *Risk Society: Towards a New Modernity,* trans. Mark Ritter. London: Sage.

Berg, Maxine. 1980. *The Machinery Question and the Making of Political Economy 1815–1848.* Cambridge: Cambridge University Press.

Binnie, Geoffrey M. 1981. *Early Victorian Water Engineers.* London: Thomas Telford.

Brante, Thomas. 1993. "Reasons for Studying Scientific and Science-Based Controversies." In *Controversial Science: From Content to Contention*, ed. Thomas Brante, Steve Fuller, and William Lynch, 177–92. New York: State University of New York Press.

Burkeman, Oliver. 2003. "Memo Exposes Bush's New Green Strategy." *Guardian*, 4 March, 16.

Collingridge, David, and Colin Reeve. 1986. *Science Speaks to Power: The Role of Experts in Policy-Making.* New York: St. Martin's Press.

Collins, Harry M., and Robert Evans. 2002. "The 3rd Wave of Science Studies." *Social Studies of Science* 32:235–96.

————. 2003. "King Canute Meets the Beach Boys: Responses to the Third Wave." *Social Studies of Science* 33:435–52.

Collins, Harry M., and Trevor J. Pinch. 1993. *The Golem: What Everyone Should Know about Science.* Cambridge: Cambridge University Press.

Conant, James B. 1951. *Science and Common Sense.* New Haven, CT: Yale University Press.

————. 1952. *Modern Science and Modern Man.* New York: Columbia University Press.

Conder, Francis R. [1868] 1983. *The Men Who Built the Railways*. Repr. of Francis R. Conder's *Personal Recollections of English Engineers*. London: Thomas Telford.

De Marchi, Bruna. 2003. "Public Participation and Risk Governance." *Science and Public Policy* 30:171–76.

Douglas, Heather. 2000. "Inductive Risk and Values in Science." *Philosophy of Science* 67:559–79.

Drangert, Jan-Olof. 1993. *Who Cares about Water? A Study of Household Water Development in Sukumaland, Tanzania*. Linköping: University of Linköping.

Edelstein, Michael. 1988. *Contaminated Communities: Social and Psychological Impacts of Residential Toxic Exposure*. Boulder: Westview.

Engelhardt, H. Tristram, Jr., and Arthur L. Caplan. 1987. "Patterns of Controversy and Closure: The Interplay of Knowledge, Values, and Political Forces." In *Scientific Controversies: Case Studies in the Resolution and Closure of Disputes in Science and Technology*, ed. H. Tristram Engelhardt Jr. and Arthur L. Caplan, 1–23. Cambridge: Cambridge University Press.

Gibbons, Michael, et al. 1994. *The New Production of Knowledge: The Dynamics of Science and Research in Contemporary Societies*. Beverly Hills: Sage.

Graham, Frank, Jr. 1970. *Since Silent Spring*. Greenwich, CT: Fawcett Crest.

Hamlin, Christopher. 1986. "Scientific Method and Expert Witnessing: Victorian Perspectives on a Modern Problem." *Social Studies in Science* 16:485–513.

———. 1988. "William Dibdin and the Idea of Biological Sewage Treatment." *Technology and Culture* 29:189–218.

———. 1990. *A Science of Impurity: Water Analysis in Nineteenth Century Britain*. Berkeley: University of California Press.

———. 1992. "Reflexivity in Technology Studies: Toward a Technology of Technology (and Science?)" *Social Studies in Science* 22:511–44.

———. 2001. "Overcoming the Myths of the North." *Forum for Applied Research and Public Policy* 16:109–14.

———. 2002. "John Sutherland's Constitutional Epidemiology." *International Journal of Epidemiology* 31:915–19.

Hamlin, Christopher, and Philip T. Shepard. 1993. *Deep Disagreement in U.S. Agriculture: Making Sense of Policy Conflict*. Boulder: Westview Press.

Harley, David. 1999. "Rhetoric and the Social Construction of Sickness and Healing." *Social History of Medicine* 12:407–35.

Herken, Gregg. 1987. *Counsels of War*. New York: Oxford University Press.

Jasanoff, Sheila. 1987. "Contested Boundaries in Policy-Relevant Science." *Social Studies of Science* 17:195–230.

———. 2000. "Reconstructing the Past, Constructing the Present: Can Science Studies and the History of Science Live Happily Ever After?" *Social Studies in Science* 30:621–31.

———. 2003a. "Breaking Waves in Science Studies: Comment on H. M. Collins and Robert Evans, "The Third Wave of Science Studies." *Social Studies of Science* 33:389–400.

———. 2003b. "(No?) Accounting for Expertise." *Science and Public Policy* 30:157–62.

———. 2003c. "Technologies of Humility: Citizen Participation in Governing Science." *Minerva* 41:223–44.

———. 2005a. *Designs on Nature: Science and Democracy in Europe and the United States.* Princeton: Princeton University Press.

———. 2005b. "Judgment under Siege: The Three-Body Problem of Expert Legitimacy." In *Democratization of Expertise? Exploring Novel Forms of Scientific Advice in Political Decision-Making,* ed. Sabine Maasen and Peter Weingart, 1–16. Dordrecht: Kluwer.

Kawachi, Ichiro, Bruce P. Kennedy, and Richard Wilkinson, eds. 1999. *The Society and Population Health Reader,* vol. 1, *Income Inequality and Health.* New York: New Press.

Kitcher, Philip. 2001. *Science, Truth, and Democracy.* Oxford: Oxford University Press.

Lambert, Richard. 1934. *The Railway King, 1800–1871.* London: George Allen and Unwin.

Latour, Bruno. 2004. *Politics of Nature: How to Bring the Sciences into a Democracy,* trans. Catherine Porter. Cambridge, MA: Harvard University Press.

Liberatore, Angela, and Silvio Funtowicz. 2003. "'Democratising' Expertise, 'Expertising' Democracy: What Does This Mean, and Why Bother?" *Science and Public Policy* 30 (3): 146–50.

Lubove, Roy. 1971. *The Professional Altruist: The Emergence of Social Work as a Career, 1830–1930.* New York: Atheneum.

Machamer, Peter, Marcello Pera, and Aristides Baltas. 2000. *Scientific Controversies: Philosophical and Historical Perspectives.* New York: Oxford University Press.

Marcus, Alan. 1985. *Agricultural Science and the Quest for Legitimacy: Farmers, Agricultural Colleges, and Experiment Stations, 1870–1910.* Ames: Iowa State University Press.

Markle, Gerald E., and Daryl E. Chubin. 1987. "Consensus Development in Biomedicine: The Liver Transplant Controversy." *Milbank Quarterly* 65:1–24.

Martin, Brian. 1996. "Sticking a Needle into Science: The Case of Polio Vaccines and the Origin of AIDS." *Social Studies of Science* 26:245–76.

Mazur, Allan. 1973. *The Dynamics of Technical Controversy.* Washington, DC: Communications Press.

Nowotny, Helga. 1975. "Controversies in Science: Remarks on the Different Modes of Production of Knowledge and Their Use." *Zeitschrift für Soziologie* 4:34–45.

———. 2003. "Democratising Expertise and Socially Robust Knowledge." *Science and Public Policy* 30:151–56.

Ottinger, Gwen. Forthcoming. "Representing Responsibility through Air Monitoring: Citizen Science, Expert Knowledge, and the Social Orders of Environmental Protection." *Social Studies of Science.*

Picart, Caroline Joan S. 1994. "Scientific Controversy as Farce: The Benveniste-Maddox Counter Trials." *Social Studies of Science* 24:7–37.

Polanyi, Michael. 1957a. "Passion and Controversy in Science." *Bulletin of the Atomic Scientists* 13:114–19.

———. 1957b. "Scientific Outlook: Its Sickness and Cure." *Science* 125:480–84.

Proctor, Robert N. 1991. *Value-Free Science? Purity and Power in Modern Knowledge.* Cambridge, MA: Harvard University Press.

Rayner, Steve. 2003. "Democracy in the Age of Assessment: Reflections on the Roles of Expertise and Democracy in Public-Sector Decision-Making." *Science and Public Policy* 30:163–70.

Richards, Evelleen, and Malcolm Ashmore. 1996. "More Sauce Please! The Politics of SSK: Neutrality, Commitment, and Beyond." *Social Studies of Science* 26:219–28.

Rip, Arie. 2003. "Constructing Expertise: In a Third Wave of Science Studies?" *Social Studies of Science* 33:419–34.

Rudner, Richard. 1953. "The Scientist *Qua* Scientist Makes Value Judgements." *Philosophy of Science* 20:1–6.

Seeley, Bruce. 1995. "SHOT: The History of Technology and Engineering Education." *Technology and Culture* 36 (4): 739–73.

Selinger, Evan, and Robert P. Crease. 2006. *The Philosophy of Expertise*. New York: Columbia University Press.

Sinclair, Bruce. 1995. "The Road to Madison and Back: Notes of a Traveler." *Technology and Culture* 36 (2, supplement): S3–S16

Suppe, Frederick. 1987. "The Limited Applicability of Agricultural Research." *Agriculture and Human Values* 4:4–14.

Toulmin, Stephen. 1982. *The Return to Cosmology: Postmodern Science and the Theology of Nature.* Berkeley: University of California Press.

Turner, Stephen. 2001. "What Is the Problem with Experts?" *Social Studies of Science* 31:123–49.

Vignoles, Keith H. 1982. *Charles Blacker Vignoles: Romantic Engineer.* Cambridge: Cambridge University Press.

Wolfe, Amy K., et al. 2002. "A Framework for Analyzing Dialogues over the Acceptability of Controversial Technologies." *Science, Technology, and Human Values* 27:134–59.

Woodhouse, Edward, et al. 2002. "Science Studies and Activism: Possibilities and Problems for Reconstructivist Agendas." *Social Studies of Science* 32:297–319.

Wynne, Brian. 1996. "Misunderstood Misunderstandings: Social Identities in the Public Uptake of Science." In *Misunderstanding Science?* ed. Alan Irwin and Brian Wynne, 19–46. Cambridge: Cambridge University Press.

———. 2002. "Risk and Environment as Legitimatory Discourses of Technology: Reflexivity Inside Out?" *Current Sociology* 50:459–77.

———. 2003. "Seasick on the Third Wave? Subverting the Hegemony of Propositionalism: Response to Collins and Evans (2002)." *Social Studies of Science* 33:401–17.

York, Herbert. 1987. *Making Weapons, Talking Peace: A Physicist's Odyssey from Hiroshima to Geneva.* New York: Basic Books.

Ziman, John. 1996. "Postacademic Science: Constructing Knowledge with Networks and Norms." *Science Studies* 9:67–80.

PART III

THE EXIGENCIES OF RESEARCH FUNDING ❖ Epistemic Values and Economic Benefit

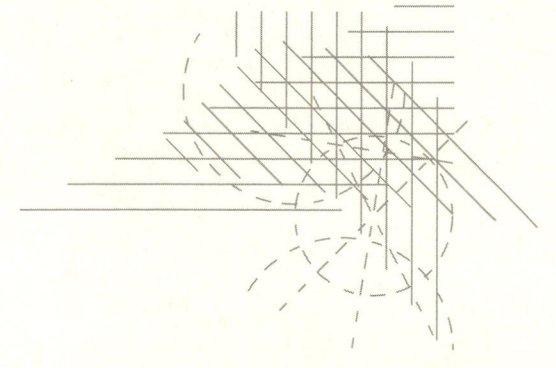

9

THE COMMUNITY OF SCIENCE®

JAMES ROBERT BROWN
University of Toronto

THE IDEA OF A COMMUNITY OF SCIENCE IS ONE we all hold dear. We think of ourselves—all academics, not just scientists in the narrow sense—as pursuing common goals and doing so in a noncompetitive way. To be sure, there are rivalries, often bitter. And no doubt we would all like the recognition that comes with being the acknowledged discoverer of something new and important. But unlike rival corporations or warring nations, our self-image is one of serving the common good—knowledge is a gift to all. Robert Merton referred to this as "communism," one of the ingredients in his famous "ethos of science" (Merton 1973).[1] Cold war prudence induced a change of name to "communalism," but the sentiment was the same—knowledge is and ought to be freely and openly shared.

Things, however, are changing. Of course, it is naive to believe there was ever a golden age when the community of science was pure and noble. But the situation is deteriorating, and it is doing so rapidly. This is because commercial interests are exacting an unprecedented influence in research. The title of this chapter has the registered trademark symbol ® attached. This is because the very phrase "Community of Science" is now the property of a consortium founded by The Johns Hopkins University designed to "accelerate the production of knowledge," as the Community of Science Inc. web site tells us.

Commercial interests are at stake, and when they are, it is both inevitable and fitting to consider a more social approach to science, though we must do so with some care. Among participants in the science wars, two of the most popular positions are little more than caricatures. One of these says that scientists adopt their beliefs exclusively on the basis of various social, political, or psychological interests. Reason and evidence are just mythic entities designed to befuddle outsiders. The other caricature of science says reason and evidence are everything. Champions of the latter view taunt their opponents with this sort of challenge: "If you think the belief that arsenic is toxic merely reflects a self-serving ideology, then you shouldn't mind taking a mouthful now." Your refusal to do so is then taken to show you are a deranged hypocrite. At the level of pure academic debate, these views are harmless. But in the public realm they can be a disaster. The one side is right to be skeptical about what is offered to us as objective research in the kinds of science that affect our lives. The financial interests of pharmaceutical companies, for instance, are unquestionably at work in how they pursue research. Yet, it is not social constructivists but rather those who believe in the possibility of reason and evidence in this domain who could actually do something about it. However, we cannot hope to make any contribution until we first realize that the anti-objectivity side is half-right—much of what passes for regular science is in reality deeply conditioned by social factors.[2] And the solution to the problem may be, at least in part, a social solution. That is, a solution will not merely involve a more rigorous application of existing methods of good science, but will also involve a social reorganization of scientific research, achieved through political action.

Arousing Suspicions

In testing the comparative efficacy of new drugs, we expect that the results could go either way. That is, when a random drug X produced by company A is compared with drug Y produced by company B, we would expect X to prove better than Y about half the time in treating some specific condition. These may indeed be the actual results of serious scientific study, but they are not the results that get published. Remarkably, when the published study in question is funded by one of the pharmaceutical companies, the sponsor's drug invariably does better. Richard Davidson (1986), for instance, in his study of 107 published papers that compared rival drugs, showed that the drug produced by the sponsor of the research was found to be superior in every single case. Lady Luck, it seems, smiles on sponsors.

The Davidson study is typical; there are many like it coming to similar conclusions, though not quite so dramatically. For instance, Friedberg et al. (1999) found that only 5 percent of published reports on new drugs that were sponsored by the developing company gave unfavorable assessments. By contrast, 38 percent of published reports were not favorable when the investigation of the same drugs was sponsored by an independent source.

Stelfox et al. (1998) studied seventy articles on calcium-channel blockers. These drugs are used to treat high blood pressure. The articles in question were judged as favorable, neutral, or critical. Their finding was that 96 percent of the authors of favorable articles had financial ties with a manufacturer of calcium-channel blockers; 60 percent of the authors of neutral articles had such ties; and only 37 percent of authors of unfavorable articles had financial ties. Incidentally, in only two of the seventy published articles was the financial connection revealed.

With these cases in mind, we should naturally become worried about who is funding the research. Whether we attribute these kinds of results to the theory-ladenness of observation, or to outright fraud, or to some new and subtle form of corruption doesn't really matter. The important questions are these: How extensive is the problem? and What are we to do about it? As for the extent of the problem, it's hard to say. The U.S. Congress passed the Bayh-Dole Act in 1980, allowing private corporations to reap the rewards of publicly funded research. Its impact has been enormous. Before Bayh-Dole there were only a couple of hundred patents each year stemming from university research in the United States. Now the annual number is in the several thousands. As all of us who live outside the United States know, American ideas and practices spread quickly. Good ideas get copied. Less than brilliant ideas are adopted, too, often courtesy of the World Trade Organization, the World Bank, or the International Monetary Fund, perhaps in the name of "free trade" or a "level playing field." Patent laws, for instance, are "harmonized," which means that U.S. patent laws must be adopted by everyone. Some of these, such as patenting organisms, have been highly significant.[3] The upshot is that commercialized medical research is forced on all.

Journal Policy

The editors of several leading biomedical journals got together to forge a common editorial policy that was published simultaneously in several journals. They had several problems to confront, but in general they were concerned with the commercialization of research and wished to protect their journals from being a "party to potential misrepresentation." In the first instance, the editors demanded full disclosure of financial relations and opposed contract research where participating investigators often do not have access to the full range of data that play a role in the final version of the submitted article. The guidelines of 2001 are part of a revised document known as "Uniform Requirements for Manuscripts Submitted to Biomedical Journals," a compendium of instructions used by many leading biomedical journals.[4] The breadth of requirements in these guidelines is considerable, from double spacing to respecting patients' rights to privacy. The editors, collectively, have added specific new requirements concerning conflict of interest. Here in outline are some main points:

- Authors must disclose any financial relations they have that might bias their work. For example: are they shareholders in the company that funded the study or manufactures the product, are they paid consultants, etc.? At the journal editor's discretion, this information would be published along with the report.

- Researchers should not enter into agreements that restrict in any way their access to the full data, nor should they be restricted in contributing to the interpretation and analysis of that data.

- Journal editors and referees should similarly avoid conflicts of interest in the peer review process.

These guidelines, if rigorously enforced, should go a long way in helping to improve the situation, and the journals should be warmly applauded for instituting them. Not only is the policy a good thing, but it was also nice to see the attention the issue received at the time in the popular media.[5] Such publicity is crucial in making the general public aware of the seriousness of the situation.

More recently, the same journal editors have taken another big step. They now require that every clinical trial be registered at the outset, that is, described in detail in some approved public form. No trial-based results would be accepted for publication unless they followed directly from a *registered* clinical trial. The point of this requirement is to prevent selective reporting and the suppression of negative results. Thus, if negative results are discovered but not published, others will at least have the opportunity to raise appropriate questions.[6]

These sorts of problems can be very serious.[7] In one troubling case, Celebrex, which is used in the treatment of arthritis, was the subject of a year-long study sponsored by its maker, Paramacia (now owned by Pfizer). The study purported to show that Celebrex caused fewer side effects than older arthritis drugs. The results were published in *JAMA* (*Journal of the American Medical Association*) along with a favorable editorial. It later turned out that the encouraging results were based on the first six months of the study. When the whole study was considered, Celebrex held no advantage over older and cheaper drugs. On learning this, the author of the favorable editorial was furious and remarked on "a level of trust that was, perhaps, broken" (quoted in Angell 2004, 109).

Selective serotonin reuptake inhibitors, known simply as SSRIs, have been central in the new generation of antidepressants. Prozac is the most famous of these. There are several drugs in the SSRI class, including fluoxetine (Prozac), paroxetine (Paxil, Seroxat), sertraline (Zoloft), and others. They are often described as miracle drugs, bringing significant relief to millions of depressed people. The basis for the claim of miraculous results is a large number of clinical trials, but closer inspection tells a different story.

There are two related issues, both connected to nonreporting of evidence from clinical trials. Whittington et al. (2004) reviewed published and unpublished data on SSRIs and compared the results. To call the findings disturbing would be an understatement. The result was favorable to fluoxetine, but not to the others. The authors summarized their findings as follows: "Data for two published trials suggest that fluoxetine has a favorable risk-benefit profile, and unpublished data lend support to this finding. Published results from one trial of paroxetine and two trials of sertraline suggest equivocal or weak positive risk-benefit profiles. However, in both cases, addition of unpublished data indicates that risks outweigh benefits. Data from unpublished trials of citalopram and venlafaxine show unfavorable risk-benefit profiles" (Whittington et al. 2004, 1341).

The related second point is illustrated in a GlaxoSmithKline internal document that was recently revealed in the *Canadian Medical Association Journal*. GlaxoSmithKline was applying to regulatory authorities for a label change approving paroxetine (Seroxat) to treat pediatric depression. The document noted that the evidence from trials were "insufficiently robust," but further remarked: "It would be commercially unacceptable to include a statement that efficacy had not been demonstrated, as this would undermine the profile of paroxetine" (quoted in Kondro and Sibbald 2004, 783). I suppose they had lots to worry about from a commercial point of view, since annual sales of Seroxat at the time were close to $5 billion.

The new journal policy requiring clinical trial registration will help put a stop to this sort of thing. It is certainly a welcome change. But not all policy changes have been happy events. Stunningly, one journal reversed its related policy on conflicts of interest. *The New England Journal of Medicine* has modified a part that previously said "authors of such articles will not have any financial interest in a company (or its competitor) that makes a product discussed in the article." The practice applies to review articles that survey and evaluate various commercial products. This class of articles is highly influential with medical practitioners. The new policy says: "authors of such articles will not have any *significant* financial interest in a company (or its competitor) that makes a product discussed in the article" (Drazen and Curfman 2002; my italics). The addition of "significant" makes quite a difference. Anything up to $10,000 is considered acceptable. Their reasons for this policy change are particularly worrisome. They think that concerns about bias shouldn't arise until significant sums are involved. Perhaps they are right. But it should be noted that someone with "insignificant" commercial connections to several different companies could be adding $50,000 to $100,000 to her income without violating the new journal rules.

The editors also claim in their editorial—and this is shocking—that it is increasingly difficult to find people to do reviews who do not have economic ties to the corporate world.[8] If they left out such reviews, they claim, they would publish nothing at all on new products, leaving readers with no means of evaluation except that provided by the manufacturers themselves. Drazen remarked that he had been able to commission only one review in the two years he then had edited the journal. The idea, one supposes, is that moderately biased information is better than none. If it is true that almost all reviewers have corporate ties—and I shudder to think it may be so—then the current situation is even worse than any reasonable paranoid would have feared.

Policy reversals seem the order of the day. Yale University once proclaimed: "It is, in general, undesirable and contrary to the best interests of medicine and the public to patent any discovery or invention applicable in the fields of public health or medicine." That was in 1948. Now Yale holds lots of patents, including one on an anti-AIDS drug. Yale shares this patent with Bristol Myers, and they enforce it to the disadvantage of the eight thousand people who die in Africa each day from AIDS because they can't pay the royalties.

Finder's Fees

The daily news is replete with horror stories. For instance, a British newspaper, the *Observer*, reports a particularly shocking case of medical abuse involving a seventy-two-year-old woman in England who was being treated by her doctor for slightly elevated blood pressure (Barnett 2003). This was a few months after her husband had died; otherwise she was in good health. Completely unknown to her, her doctor enrolled her in a clinical trial, gave her various pills that had serious side effects, and regularly took blood to the point that her arms were "black and blue." Some of the pills were given directly by the doctor, not through the usual process of taking a prescription to the pharmacy. Her suspicions were aroused and after a particularly bad reaction to one pill, she complained to health authorities. The subsequent investigation revealed that her doctor had been given £100,000 over the previous five years for enrolling patients in clinical trials at £1000 each. The companies involved include AstraZeneca, GlaxoSmithKline, and Bayer. Many of the doctor's patients did not know they were being enrolled, many did not have any of the relevant symptoms to be included in the study in the first place, and many patients who did have relevant symptoms were given placebos, instead of the standard treatment they required.

One might hope that this culprit is just an isolated bad apple. But when we hear that in the United Kingdom more than three thousand GPs are enrolling patients in clinical trials at £1000 each and that the pharmaceutical industry is spending more than £45 million for patient recruitment in the UK, then it is not such a surprise to learn that there are dozens of examples of fraud. In the case of one GP, the consent forms of twenty-five of the thirty-six patients he enrolled were forged. Another who collected £200,000 failed to notify patients of possible side effects. He was subsequently caught offering a bribe to one of those patients not to testify (Barnett 2003).

If this much corruption can be generated by a finder's fee of £1000, imagine what might happen if the fee were tripled. In the United States in 2001 the average bounty was $7000 per patient (Angell 2004, 31).

Payment for recruiting is known as a "finder's fee." But the term is hardly ever used, since the idea is often thought to be unacceptable. The fee is usually hidden in so-called "administrative costs" or perhaps disguised in some other

way for which compensation is considered acceptable, such as well-paid consul-
tantships or invitations to conferences held in exotic and luxurious settings. In
any case, the recruitment can be so profitable that one family practice organization in the United States placed an ad on the Internet: "Looking for Trials! We
are a large family practice office. . . . We have two full time coordinators and
a computerized patient data base of 40,000 patients. . . . We are looking for
Phase 2–Phase 4 trials as well as postmarketing studies. We can actively recruit
patients for any study that can be conducted in the Family Practice setting"
(quoted in Lemmens and Miller 2003).

There are all sorts of concerns that arise with recruiting. Most of these are
ethical issues. Since my focus is on epistemology, I am not concerned with them
here but will mention a few in passing.

- Outright fraud. This arises, for example, from forged consent forms.

- Lack of treatment. Some subjects with treatable conditions will be put
  into the control group and given placebos. Existing treatments, from
  which they could benefit, will be bypassed.

- Safety. People who have health conditions that make them inappropriate for a particular study are being included because the financial
  incentives for inclusion are so great.

- Privacy. Health records are gathered over the Internet to build large
  commercial data bases that are not particularly secure.

My interests, as I said, are epistemic. Here are some of the methodological
problems that arise from an epistemic point of view.

- It is increasingly difficult to find test subjects for government-
  sponsored research, since typically no finder's fee is offered. This
  means that it is becoming more difficult to do research that is relatively
  independent of economic interests.

- Incompetence. This arises when GPs, for instance, become involved in
  clinical trials; they have no particular skill or training in research.

- Improper inclusion. The criteria for inclusion are improperly
  expanded so as to make it easier to recruit test subjects. This weakens
  the reliability of experimental results.

Critics of commercialized research tend to focus on moral improprieties, such
as a lack of informed consent or a failure to administer effective known treat-

ments. However, even if these moral requirements are breached, the subsequent science might still be perfectly good from an epistemic point of view. The research itself might, of course, suffer in cases where recruited subjects failed to fit the protocol, that is, they do not have the relevant health condition. This sort of case involves both moral and epistemic failings. Many of these moral and epistemic problems can be controlled, at least in principle, by regulation. Conflict of interest rules should be able to prevent abuses of the sort I mentioned, though not without difficulty; recall that finder's fees can be hidden in so-called administrative costs, thus concealing the otherwise evident conflict. In any case, most discussions of these issues focus on regulating conflict of interest.[9] But an additional epistemic problem arises that is independent of these considerations and cannot be controlled by the same sorts of conflict of interest regulations. The problems I have been describing so far are sins of commission. The problem I will presently describe is more like a sin of omission. It is a huge epistemic problem, and it concerns a lack of alternative theories.

Skewing Research toward the Patentable

This point is so obvious, it hardly needs to be mentioned. Yet it is of prime importance. Corporations understandably want a return on their investments. The payoff for research comes from the royalties generated by exclusive control of intellectual property. This means corporations will tend to fund only research that could in principle result in a patent. Other kinds of information are financially useless.

Imagine two ways of approaching a health problem. One way involves the development of a new drug. The other way focuses on, say, diet, exercise, or environmental factors. The second could well be a far superior treatment, both cheaper and more beneficial. But obviously it will not be funded by corporate sponsors, since there is not a penny to be made from the unpatentable research results. It should be just as obvious that a source of funding that does not have a stake in the outcome but simply wants to know how best to treat a human ailment would, in principle, fund either or both approaches, caring only to discover which is superior.

To get a sense of what is at issue here, consider a comparative trial carried out on patients who were at high risk of developing diabetes. Over a three-year period, 29 percent of the placebo group went on to develop diabetes; 22 percent who took the drug metformin developed diabetes; but only 14 percent of those who went on a diet and exercise program developed the disease (Angell 2004,

170). Clearly, the best result came from a treatment that is not patentable. This trial, by the way, was sponsored by the U.S. National Institute of Health, not by commercial interests.

In a study of the effects of exercise on depression, Dunn and coresearchers found significant results. "In summary, aerobic exercise in the amount recommended by consensus public health recommendations was effective in treating mild to moderate MDD [major depressive disorder]. The amount of exercise that is less than half of these recommendations was not effective. Rates of response and remission with a PHD [public health dose, commonly recommended amount] are comparable to the rates reported in trials of cognitive behavioral therapy, antidepressant medication, and other exercise studies" (Dunn et al. 2005, 7). These are valuable findings, but they are not patentable.

Of course, there is also the problem that members of the public often want a quick solution and are not that keen on diet and exercise. Yet, if they were not so bombarded with industry propaganda, or if they got an equal amount of publicly sponsored information about the relative benefits of diet and exercise (perhaps presented in a humorous way like the Viagra ads), then we might well see more people opting for the better solution. Public funding is clearly the answer to several aspects of this epistemic problem.

Even within patentable research, some areas will be less profitable than others. Consequently, diseases of the poor and the developing world (for example, malaria) have gone relatively unexplored, since the poor cannot afford to pay high royalties. We are also in danger of losing a genuine resource in the form of top-notch researchers who do not do patentable work. In an example outside medical research, the University of California at Berkeley formerly had a Division of Biological Control and a Department of Plant Pathology, but neither exists today (Press and Washburn 2000). Why? Some people close to the scene speculate that it is simply because the type of work done in these units is not profitable. Typical research in these units involved the study of natural organisms in their environments carried out with a view to controlling other natural organisms. This type of work cannot be patented. Is it valuable? Yes. Is it profitable? No.

Trends being the way they are, top graduate students will not go into the field. Fewer and fewer people will work on agricultural and environmental problems through biological control. Perhaps the petrochemical industry will be able to solve all our agricultural problems. It is not the job of a philosopher to speculate on this possibility, but it is the job of philosophy of science to make the methodological point that without seriously funded rival approaches, we will never know how good or bad particular patentable solutions really are. The epistemic point is a commonplace among philosophers. Evaluation is a *comparative* process. The different background assumptions of rival theories lead us to see the world in different ways. Rival research programs can be compared in terms of their relative success in the long run. But to do this, we need strong rivals for the purposes of comparison.

Deliberate Ignorance

There is considerable evidence that the tobacco industry took legal advice to the effect that they should *not* do any research into the possible harmful effects of tobacco. Had they come to know of any harmful effects, their legal liability would have greatly increased. When such information came into their hands, they tried to suppress it. Obviously, from their viewpoint, it is better not to know about it in the first place. Given the potential for lawsuits over liability, ignorance is bliss. Interestingly, many of the lawyers who advised the tobacco industry also advise pharmaceutical companies. The legal firm of Shook, Hardy & Bacon, for instance, advises both the tobacco industry and the pharmaceutical firm Eli Lilly (Schulman 1999).

This legal strategy assumes a distinction philosophers know well—the distinction between discovery and justification. The thinking seems to be that a vague hunch that tobacco causes lung cancer or that SSRIs sometimes induce suicide are just that—vague hunches. And hunches do not constitute evidence. For that we would need extensive clinical trials. Since these trials have not been done, we simply have no evidence at all, according to the relevant corporations. There is no justification, according to them, for making these claims about the harmfulness of tobacco or SSRIs.

There is a great deal of naiveté about scientific method—sometimes amounting to willful ignorance. Some researchers claim that clinical trials are both necessary and sufficient for definite knowledge and that anything short of a full clinical trial is useless. (Champions of so-called Evidence Based Medicine occasionally make this claim.) This all or nothing attitude is ridiculous. We choose which clinical trials to run on the basis of plausibility; circumstantial and anecdotal considerations play a decisive role. This, too, is evidence, though usually not as strong. Often, though, these plausibility considerations are enough, or should be enough, to launch a serious study. Refusing to take action on the grounds that there is nothing but "anecdotal evidence" is not only bad philosophy of science, it can also be criminal. Even the Nazis clearly established the link between smoking and lung cancer in the 1930s (Proctor 1999). Commercial interests elsewhere stood in the way for more than two decades, during which time millions of people died.

I mentioned SSRIs above. Selective serotonin reuptake inhibitors are widely used for combating depression. Prozac is perhaps the best known of these. There is a lot of interesting stuff to be discovered here. It may be that SSRIs actually improve depressed people's condition to the point of suicide. This sounds paradoxical. What seems to happen is that extremely depressed people are sometimes in a "nonresponsive or lethargic" state. The SSRI will improve their condition to the point where they have the energy and the wherewithal to commit suicide. At the other end of the spectrum, even some healthy nondepressed volunteers have become suicidal after taking SSRIs. Needless to say, this is something a profit-seeking corporation is reluctant to investigate.

David Healy is a British psychiatrist, currently the director of the North Wales Department of Psychological Medicine at the University of Wales. He has spoken and written extensively on mental illness, especially on pharmaceuticals and their history (see, for example, Healy 2001). In September 2000, he was offered a position as director of the Mood and Anxiety Disorders Program in the Centre for Addiction and Mental Health (CAMH) affiliated with the University of Toronto. As part of the deal, Healy was also appointed as professor in the university's Department of Psychiatry. He accepted this appointment. Before he moved permanently to Canada, he took part in a Toronto conference in November 2000, where he gave a talk that was quite critical of the pharmaceutical industry. Among other things, he claimed that Prozac and other SSRIs can cause suicides. Within days of his talk at Toronto, CAMH withdrew the appointment. Healy was, in effect, fired before he started.

Needless to say, this has been quite a scandal. What happened? There are, of course, rival views. The official account coming from CAMH is that they realized they had made a mistake, that Healy would not be a suitable appointment on purely academic grounds. Another view suggests that pressure from Eli Lilly did him in. Neither of these seems very plausible. Much more believable is the view that Lilly did not put any pressure on, but that self-censorship was at play. Lilly contributes financially to CAMH, a fact that lends credence to this last speculation. There would be no need for direct pressure, if the recipients of Lilly's largess are ever ready to take the initiative themselves. This need not be conscious; some values are easily internalized.

More recently there have been very serious charges that drug companies have tried to suppress data on the harmful effects of SSRIs on children. I described this above in connection with journal policies requiring full disclosure. It is a clear vindication of Healy's concerns. But we need not worry about the details of the Healy case. The crucial thing is that it is not in the interest of Eli Lilly or other pharmaceutical companies to "know" that Prozac or other SSRI products cause suicide, since that would increase their potential liability. If work is to be done on this issue, it will have to be publicly funded. Lilly is not likely to foot the bill.

Creating a New Disease

Let us consider a different type of example. Eli Lilly has recently promoted a product called Sarafem for those who suffer from premenstrual dysphoric disorder, or PMDD. PMDD is an updated version of PMS, premenstrual syndrome. This alleged mental disorder is said to affect some women in the luteal phase of the menstrual cycle, just before the onset of menses. The American Psychiatric Association has not yet accepted that PMDD is a disorder, but it does list it in the appendix of the bible of this field, *Diagnostic and Statistical Manual of Mental Disorders*, known in its latest version as DSM-IV. To be diagnosed as having PMDD, a woman must have five or more of the following eleven symptoms, which characterize the disorder:

- markedly depressed mood
- marked anxiety
- marked affectivity
- decreased interest in activities
- feeling sad, hopeless, self-deprecating
- feeling tense, anxious, or "on edge"
- persistent irritability, anger, and increased interpersonal conflicts
- feeling fatigued, lethargic or lacking in energy
- marked changes in appetite

- a subjective feeling of being overwhelmed or out of control

- physical symptoms such as breast tenderness, swelling or bloating

Frankly, I don't know anyone—male or female—who fails to satisfy at least five of these conditions from time to time. In any case, Sarafem's active ingredient is fluoxetine hydrochloride, the same active ingredient as the antidepressant Prozac, which is also made by Eli Lilly. However, Lilly is definitely promoting Sarafem in a different way. Marketing associate Laura Miller said, "We asked women and physicians about the treatment of PMDD, and they told us they wanted a treatment option with its own identity that would differentiate PMDD from depression. PMDD is not depression. As you know, Prozac is one of the best known trademarks in the pharmaceutical industry and is closely associated with depression. They wanted a treatment option with its own identity" (quoted in O'Meara 2001).

What then is the difference between taking a dose of Prozac and a dose of Sarafem? In either case, it is 20 mg of fluoxetine hydrochloride (though the pills have changed color from green to lavender—surprised?). "The difference," according to Miller, "is that PMDD is a distinct clinical condition different than depression. PMDD is not depression. PMDD is cyclical—women suffer from PMDD up to two weeks before their menses, and the other two weeks of the month they don't have the symptoms of PMDD" (O'Meara 2001).

Lilly was about to lose much of its patent protection on Prozac. It is hard to resist speculation on the connection to the promotion of PMDD and Sarafem. Patent laws will protect a discovery if it is a distinct new use of an already existing entity. For patent protection, then, it was crucial that Lilly find a new use for fluoxetine hydrochloride. If PMDD is depression, they are out of luck.

Once again, the philosophical moral is evident. Through clever marketing, advertising, and public relations, Lilly (in some sense), is creating a disease. If they can first sell the psychiatric illness, they can then sell the cure. How this takes place is an enormously curious and philosophically interesting thing. I think it is safe to say that PMDD did not exist in the past, but it might start to exist in the near future.

Let me explain. The tame sense of saying Lilly is "creating" a disease is the sense in which they are merely getting us to believe that such a disease exists. There is no doubt that they are trying to do this. But there may be a deeper sense in which they are creating the disease. Ian Hacking (1995, 1998) has written extensively on so-called "transient mental illness," which is not just transient in an individual, but in a society. Multiple personalities, mad travelers, and anorexia are likely examples (though debatable, of course). This type of mental illness comes rather suddenly and spontaneously into existence at a specific time and place, and just as quickly disappears. Taking a cue from this, there are three interesting possibilities for PMDD.

- The disease PMDD has always existed. Lilly is merely bringing to light a fact that their research has uncovered and is promoting Sarafem as a way to treat it.

- The disease does not exist. Lilly, however, is trying to get us to believe that it does, anyway, since that will lead to sales of Sarafem.

- The disease has not existed in the past, but the public relations activities of Lilly will create the disease (perhaps like other transient mental illnesses), and that will lead to sales of Sarafem.

In the first case we should be thankful to Lilly, since a problem will have been brought to our attention and a remedy provided. If the second or third possibility is actually the case, then the dangers of private funding of market-oriented medical research are manifestly clear.

There are a number of additional topics I have not even touched upon. These provide even more reason for serious concern with current medical research practice. For one thing, vaccine research has declined, since vaccines are nowhere near as profitable as drugs for chronic conditions. For another, respectable medical journals, in order to help finance themselves, allow special supplemental issues. These are often little more than advertising outlets for the corporate sponsors, but they have the same format as the regular peer-reviewed issues, so readers are easily misled. The list goes on.

The Bold Entrepreneur

We are all familiar with the popular image of the entrepreneur, the bold and innovative risk taker, whose initiatives benefit us all. Well, it is certainly true that some have benefited. The pharmaceutical industry in the United States does well over $200 billion a year in business. Profit levels are at a staggering 18 percent of sales, the highest of any U.S. industry listed in the Fortune 500. (The median of that group has profits of less than 4 percent of sales.) How could we account for this extraordinary success? Does it have something to do with the 11 percent of the $200 billion that goes into research? That is certainly a lot of money. Or does it have more to do with the 36 percent that goes into marketing? That is more than triple the research budget.

In 2002 the U.S. Food and Drug Administration (FDA) approved seventy-eight new drugs. Only seven were classified as improvements over older drugs. The rest are copies, so-called "me too" drugs. Not one of these seven was produced by a major U.S. drug company. There is nothing special about the year 2002. During the period 1998–2002 the FDA approved 415 new drugs and classified them as follows:

- 14 percent were new innovations;

- 9 percent were significantly improved old drugs;

- 77 percent were no better than existing drugs.[10]

The last of these are the "me too" drugs. They are copies of existing drugs (not exact copies, since they have to be different enough to be patentable). By U.S. law, the FDA must grant approval so long as a new drug is "effective," which just means that it does better than a placebo in a clinical trial. Once approved, marketing takes over and monstrous profits can be made. A cheaper and better generic alternative will not be similarly promoted, since there is no hope of stupendous earnings.

Not surprisingly, there are calls for clinical tests to compare new drugs with the best existing alternative, not just with placebos. The importance of such comparative testing is illustrated by a massive comparative study on various types of blood pressure medication. It was carried out by the National Heart, Lung, and Blood Institute (part of NIH) and was almost completely publicly funded. The result was striking: the best medication turned out to be an old diuretic ("water pill"); it worked as well or better than the others, had considerably fewer side effects, and costs about $37 per year, compared with several hundred dollars annually for each of the others (Angell 2004, 95).

It is clear that licensing should then be based on relative performance. This is the sort of problem that could be obviously and easily controlled with proper regulation. It is hard to imagine a more poorly constructed regulatory system than one currently in place in the United States. It leads one to think that U.S. lawmakers are either a pack of fools or as corrupt as the pharmaceutical companies who lavishly lobby them. There is ample evidence for either conclusion.

Where does genuine innovation come from? Consider the case of Taxol (paclitaxel). It is a very important drug, widely used to treat various forms of cancer. It was derived from the bark of the Pacific yew tree in the 1960s. The cost of this research was $183 million, paid for by taxpayers through the National Cancer Institute. However, in 1991 Bristol-Myers Squibb signed an agreement with NCI. The upshot is that they, not taxpayers, make several millions in royalties each year on annual sales of up to $2 billion (Angell 2004, 58). This is a common pattern: Risk and innovation are paid for by the public purse; profits are privatized. How profitable can this be? When Taxol was brought to market, the cost of a year's treatment was $10,000 to $20,000, a tenfold increase over the cost of production (Angell 2004, 66). It is hard to know whether we should feel outrage or admiration. Imagine getting the public to pay for this research not once, but twice.

The swaggering entrepreneurs of the pharmaceutical industry boast that they are doing risky, innovative research. This is pure nonsense. The only innovative business they are in is marketing. They are utterly dependent on the advice of their scientists; their only expertise lies in the skilled promotion of their products. Even if we admired the genuine entrepreneurial spirit in principle, it is only a joke here. They are nothing like Alexander Graham Bell inventing the telephone, Marconi the radio, or even Bill Gates developing software for the home computer. Big Pharma consists of business people exploiting the intellectual work of others. Frankly, who needs these parasites? Their only contribution to medicine has been marketing, and it is far from clear that the public has benefited from being told they may be suffering from PMDD or erectile dysfunction. On the other hand, the public has certainly been hurt by the suppression of discoveries of harmful side effects and by financial decisions that set prices at the highest level the market will bear. People suffer, and people die.

If the "me too" drug dealers do not quite fit our image of the heroic innovator, perhaps they can live up to their other highly touted virtue—efficiency, a principle part of the "magic of the market." Could it be that publicly funded research cannot hope to compete with the highly efficient private organizations? I will say more on this below.

In any case, one has to wonder to what extent the combination of entrepreneurial medical research and the promotion of creationism in its schools is turning the United States into a scientific backwater. There are many strong voices opposing both trends, but unless the apathetic majority of U.S. scientists takes up the cudgels, further degradation of a once great scientific establishment is all we can expect.

Epistemology and Science Policy

Epistemology has always had a normative component. We *ought* to accept the theory that provides the best explanation; we *ought not* to accept any theory that is inconsistent; and so on. Typically, however, philosophers of science stay away from public policy. Probably we all would just smile if our political leaders were to ask: "What should our policy on patents be, if the Bayesians win out over the Popperians?" or if the Minister of Science and Technology should inquire: "In light of the failure of Hempel's account of explanation, how much money should we put into high-energy physics?" We don't think of it as our job to tell governments how to organize science. Policy decisions must be based on social goals and other factors that have nothing to do with epistemology.

However, the considerations we have described so far strongly suggest that such a hands-off attitude is insupportable. Those of us interested in the epistemology of science have to get involved in science policy, unless we mean epistemology merely to be a descriptive enterprise, perhaps in the spirit of SSK (the sociology of scientific knowledge). The social constructivist David Bloor, for instance, thinks of himself as describing but never prescribing anything to do with science (Bloor [1976] 1991). By contrast, it is those of us who believe in genuine objective norms for science who can have something meaningful to propose. There is such a thing as objective science, we would say, and its canons are being seriously violated in current medical research *because of the way research is socially organized*. Policy pronouncements can and should be made by those with a concern for good methodology. This advice need not be at a very detailed level, but neither would it amount to merely uttering a few pious platitudes about the need for intellectual honesty.

Steven Shapin, a well-known historian of science, writes amusingly about the reliability of observers in the context of seventeenth-century English science (Shapin 1994). The prevailing opinion was that servants and women were not to be trusted to tell the truth, since they were not appropriately independent. Nor were Catholics, since they could be under the influence of the Jesuit doctrine of "mental reservation." This doctrine allowed English Catholics after the Reformation to avoid telling a lie by mentally adding "It is certainly false that" to their voiced avowal "I am a Protestant." It may be a neat trick for avoiding religious persecution, but it is disconcerting for collectors of empirical data who rely on the reports of others. The ideal observer is, as you might imagine, the English gentleman: rich, independent, and possessing sufficient moral fiber as to resist all corruption. His observations, and his alone, we can trust. There are not, alas, enough English gentlemen around to serve the considerable needs of contemporary science. However, plan B is waiting in the wings. The next best thing is to guarantee the independence of the researchers by doing a few simple but significant things. I will confine my suggestions to medical research, since that is scope enough and I quite deliberately want to associate my view with existing social policy in most countries where some sort of national health service is in existence. But first, let us consider the available research options. I see three.

1. Free market

2. Regulated market

3. Socialized research

The first and second options share the view that medical research should be conducted on a market basis. They differ on how regulated this market should be. The difference between them is a matter of degree, but it could be considerable. Champions of the first view would, of course, allow that fraud should be outlawed, but after that they would want as little regulation as possible. Champions of the second view could conceivably want to instill very strong regulations.

The following are the kinds of regulations I have in mind. I have mentioned several possible regulations already, but they are worth repeating, briefly. Of course, would-be regulators would likely not agree with all of these, but they include the kinds of things that champions of a regulated market might consider:

- Require full disclosure of any financial interest when publishing

- Require advance registration of any clinical trial as a condition of publishing, or better yet, establish an independent agency to design, conduct, and interpret all clinical trials

- Require clinical trials to test products against leading alternatives, as well as against placebos

- Disallow finder's fees (including disguised equivalents)

- Disallow corporate-sponsored "education" of doctors

- Disallow public marketing of drugs

Regulations such as these could be instituted to control some of the problems of market-driven medicine. But there are still shortcomings in any market model of research, even with massive regulation. For one thing, it is almost impossible to get the regulations to cover all the serious cases. The rules must be fleshed out in considerable detail in the hope of anticipating all serious problems. The first, involving financial disclosure, for instance, needs to cover not just the obvious direct cases, but indirect ones, too. For example, a prominent cardiologist who strongly criticized Vioxx turns out to have been a consultant for a hedge fund that was betting that the stock of Merck (the maker of Vioxx) would fall (Pollack 2005). Perhaps he was not influenced, but the potential for problems are glaring. Trying to anticipate the full scope for corruption is almost hopeless. Even when anticipated, how good will enforcement be? As they currently stand, conflict-of-interest policies instituted by medical journals are largely based on an honor system. Sheldon Krimsky's preliminary investigation (2003, 199) suggests that compliance is far from ideal.

Though I doubt regulation will handle all the problems that prevail in medical research, licensing, and so forth, some regulation will be required in any system. Rules with teeth and government agencies that operate at arm's length are essential. Such agencies must be free of any sort of governmental or industry influence. This is not currently the case in the United States. David Willman, an investigative journalist for the *Los Angeles Times*, has uncovered a number of troubling cases:

- Dr. P. Trey Sunderland III, a senior psychiatric researcher, took $508,050 in fees and related income from Pfizer Inc. at the same time that he collaborated with Pfizer—in his government capacity—in studying patients with Alzheimer's disease. Without declaring his affiliation with the company, Sunderland endorsed the use of an Alzheimer's drug marketed by Pfizer during a nationally televised presentation at the NIH in 2003.

- Dr. Lance A. Liotta, a laboratory director at the National Cancer Institute, was working in his official capacity with a company trying to develop an ovarian cancer test. He then took $70,000 as a consultant to the company's rival. Development of the cancer test stalled, prompting a complaint from the company. The NIH backed Liotta.

- Dr. Harvey G. Klein, the NIH's top blood transfusion expert, accepted $240,200 in fees and 76,000 stock options over the last five years from companies developing blood-related products. During the same period, he wrote or spoke out about the usefulness of such products without publicly declaring his company ties (Willman 2003).

It seems unnecessary to comment further on these three examples. Sometimes facts really do speak for themselves.

Yet another episode will help to undermine any remaining confidence in U.S. regulatory agencies. The FDA suspended marketing of a number of pain-killers following the release of data suggesting seriously harmful side effects. These included Celebrex, Bextra, and Vioxx (known as Cox-2 inhibitors). The decision was reviewed by a committee of 32 government drug advisers. They voted to endorse continued marketing. It turns out that 10 members had financial ties to the drug makers. If they had excused themselves on grounds of conflict of interest, the outcome would have been quite different. Had they not voted, two of the drugs would not have been reinstated (Celebrex would have been reinstated either way). The votes would have been 12 to 8 against Bextra and 14 to 8 against Vioxx. The 10 members with ties to the drug makers voted 9 to 1 in favor in each case (Harris and Berenson 2005).

Getting good regulations in place is vitally important. Nevertheless, the problems that regulation actually solves tend to be the lesser ones, and often they can be solved in a different way. More serious is the problem that there is no incentive to do research on medical solutions to health problems that cannot be patented. It is the crucial generation of a wide class of rival theories that is totally lacking in for-profit research. And that is my main reason for preferring the third option, socialized research. I recommend the following actions:

- Eliminate patents in the domain of medical research
- Adjust public funding to appropriate levels

What can be said in favor of these two points? It might be thought that patents are necessary to motivate brilliant work. Nonsense. The most brilliant work around in mathematics, high energy physics, and evolutionary biology is all patent-free. Curiosity, good salaries, and peer recognition are motivation enough. What about the problem that a great deal of medical research is simply drudge work, namely, massive clinical trials? This may be true, but clinical trials are going to be needed for some types of research that are clearly not patentable and just as clearly are of great use to society. If public funding works for clinical trials for the influence of broccoli on health, where nothing is patentable, then public funding can work for drugs, too.

Why do I call for "appropriate" levels of funding rather than for matching current levels? For one thing, it is hard to tell what current levels are. Drug companies claim that it costs on average more than $800 million to bring a new drug to market. This, however, is a gross exaggeration. Something like $100 million is a more reasonable estimate, since marketing costs (which they include) are not part of genuine research.[11] Moreover, many research projects are for "me too" drugs, which bring little or no benefit to the public. When we take these factors into account, it is clear that we can maintain a very high level of research for considerably less public money.

Can these proposals actually be implemented? Methodological issues are tough enough. Policy is vastly more messy, since it must be instituted in a social context. Radical proposals have little chance of success, even if they are impeccable from a methodological point of view. My proposal to socialize medical research may seem quite radical, but it actually is not, at least not in many societies. In Canada and other countries with socialized medicine, the attitude of the general public is that medicine is one place where the market should not rule. (Education is another area where, at least at lower levels, there is near universal support for free publicly funded schooling for all.) In this context a proposal such as mine fits seamlessly into the existing national health care system. If anything, private enterprise medical research is the oddity. Needless to say, this is not true in the United States, but elsewhere in the industrialized world it is. This means that as a policy, it should be relatively easy to implement. In short, socialized research goes hand in hand with socialized medicine.[12]

But isn't this an invitation to government waste? Won't free enterprise research be more efficient in every respect? After all, isn't it a well-known fact that socialists, however well-meaning, are bad with money and hopelessly inefficient? Since I'm tying medical research to socialized medicine, we would do well to compare the relative efficiency of two types of health care. Woolhandler et al. (2003) compared the cost of health care administration in the United States and Canada. They concluded: "In 1999, health administration costs totaled at least $294.3 billion in the United States, or $1,059 per capita, as compared with $307 per capita in Canada." They noted that the spread was getting worse and draw the obvious moral. "The gap between U.S. and Canadian spending on health care administration has grown to $752 per capita. A large sum might be saved in the United States if administrative costs could be trimmed by implementing a Canadian-style health care system" (Woolhandler et al. 2003, 768).

The comparison is even more stunning when examining the costs of the two systems as percentages of GNP. Health care in the United States costs 15 percent of GNP and yet leaves roughly a quarter of the population uncovered. In Canada the cost is less than 10 percent of GNP and everyone is covered.

So much for socialist inefficiency. Of course, one needs a government that is committed to being efficient. Even the most pro-market political parties in Canada realize that there is no going back on socialized medicine in Canada (or so they say), and large corporations support it, since it keeps their costs considerably lower than their U.S. competitors. So it is in the government's interest as well as in the general public's to run things efficiently. By contrast, some politicians do not want any government agency to run efficiently, as that would be a challenge to the private sector. Consider the politicians who took the following action: Medicare, a U.S. health program for senior citizens, was prohibited by a law passed by the U.S. Congress from using its potential buying power to bargain for lower prices. That government agency must pay top price. Hands up, all those who think this law was passed by socialists.

Values and Methodology

There are a handful of human activities that are completely ennobling. The list is no doubt headed by anything that alleviates poverty and suffering. It also includes the production of great art and great science. Medical research should be near the top of this list. Yet all that is wonderful and noble is corrupted by commerce and degraded by greed. Half a century ago, Jonas Salk discovered the polio vaccine. He prevented millions of deaths and millions of hearts from breaking. When asked if he would patent his finding, he replied: "There is no patent. Can you patent the sun?" We can't all make the sort of contribution Salk made, but we can, each and every one of us, avoid the slimy swamp.

It might seem that my insistence on eliminating patents in medical research, on being involved in policy decisions, and on the particular policy advocated, is a mere reflection of various values that I happen to hold. Perhaps this is so. Indeed, I am sure that it is. But there is another way to consider this issue, a way in which values do not play any determining role at all. In fact, I consider the whole business a question of good methodology, not morals.

Scientific method is not fixed for all time but rather seems to evolve, often under the influence of scientific discoveries. For instance, the discovery of placebo effects led to the introduction of blind and double-blind tests. That is, the discovery of a fact led to the institution of a norm: In situations of such and such a type, you *ought* to use blind tests. Though this is a norm, it is not ordinarily what would be called a value. That is, it is not a social or moral value, though it is certainly an epistemic value. The practice of blind testing, as others have noted, is best seen as science simply adopting what it has itself established as appropriate methodology. I suggest that we can view the current situation in a similar light. We have learned empirically that research sponsored by commercial interests leads to serious problems, so serious that the quality of that research is severely degraded. This was the point of my citing so many examples. The switch to public funding solves many, if not all, of these epistemic problems. Therefore, as an epistemic norm we should have public funding for medical research. This line of reasoning is no different than, first, discovering the placebo effect, next, discovering that blind testing can overcome the difficulties that the placebo effect entails, and then, as a result of all this, adopting the methodological norm of employing blind tests.

Let me take a moment to anticipate a possible objection. If the case I made for eliminating intellectual property rights holds in medical research, shouldn't it hold in general? And if it does, shouldn't patents everywhere be eliminated? Or, to put it the other way around, if it's a bad idea to eliminate them everywhere, then it must be a bad idea to eliminate them in medicine. My reply is simple: I deny the universality of the argument. When it comes to the economy, few people today take the all-or-nothing view that the government should own all the means of production or none whatsoever. The most successful economies are mixed. Some industries and institutions (trains, schools) are best run publicly, while others (restaurants, clothing) are best left to free enterprise. Trial and error is the sensible thing. I take the same attitude with patents. I'm sure they are very beneficial to society in some areas. But medicine is not one of them.

Scientific Socialism

The expression "scientific socialism" might be taken as a fair description of the view of medical research I have been urging. But those who have some acquaintance with Marxism may conjure up a different image. Marx and Engels famously distinguished their outlook from utopian socialism. The difference is quite important. The motivation for utopian socialism is primarily moral. It is promoted by those who are outraged—as they should be—by poverty and social injustice. Social change, if it comes, would be an ethical response to the horrors of the prevailing situation. The scientific socialism of Marx and Engels takes quite a different view. Capitalism, according to them, contains the seeds of its own destruction. Capitalist competition leads inevitably to the concentration and impoverishment of the working class. Socialism will grow out of this state of affairs in a perfectly natural and inevitable way. Though morally superior, socialism, according to any traditional Marxist, is not the result of moral choice, but rather the outcome of an inevitable historical sequence, the result of something akin to a law-like process.

I see the problems involved in medical research in a somewhat similar way. This is, to repeat, quite different than advocating public funding for moral reasons. The policy I urge is motivated by epistemology. The contrast is not unlike the contrast between utopian and scientific socialism. It is not moral outrage—though I certainly acknowledge its presence—but rather the internal logic of capitalism (according to Marx) or of current medical research (according to me) that drives things along.

There is, however, one glaring disanalogy. According to Marx's scientific socialism, the internal logic of capitalism leads inevitably to socialism. However, the internal logic of current medical research does not lead inevitably to socialized research. It still requires political action to bring it about. And that is anything but inevitable. It is at this point that philosophical argument in the public domain becomes essential. And here I would expect nothing less than a long, hard battle.

Wilfrid Sellars once remarked that philosophy is a Dedekind cut between the introductory remarks and the conclusion. In these terms my introduction was the long list of problems that arise from the commercialization of medical research. My conclusion was a call for the elimination of patents in medicine. The Dedekind cut was that bit that went by in a flash distinguishing moral outrage from methodological fine tuning. It might be worth repeating.

Advocating public funding for research in the health sciences might appear to stem from one's socialist ideology. This is so in part, but it is very much more than that. The need for patent-free public funding in medical research is like the need for blind testing. It is, to put it simply, an epistemic discovery made by many people. Facts have been uncovered that require a methodological response, not a moral one. The right response, I urge, is to socialize medical research. The fact that scientific socialism, as I am here calling it, harmonizes well with one's moral sense, at least for me, is a happy accident.

NOTES

1. There are four ingredients in Merton's ethos of science: universalism, communism, disinterestedness, and organized skepticism (1973, 270).

2. For a discussion of these issues, see my earlier book, *Who Rules in Science? An Opinionated Guide to the Wars* (2001).

3. In 1980, the same year as the Bayh-Dole Act, the U.S. Supreme Court ruled in favor of patenting living organisms in *Diamond vs. Chakrabarty*. The 5-4 decision shows how contentious the issue was. The organism in question is the microorganism *Pseudomonas*, useful for cleaning up oil spills. The decision opened the door to patenting living things in general, and in 1988 the Harvard OncoMouse was patented. Interestingly, the Supreme Court of Canada has bucked this trend and refused to allow patents on the Harvard mouse.

4. It can be found, for instance, in *Lancet* 358 (15 September 2001): 854–56. The same document is also in *JAMA*, *New England Journal of Medicine*, and in many other journals in issues published at roughly the same time in mid-September 2001. It is also available online at http://www.icmje.org/index.html.

5. For example, see "Medical Journals to Set New Policy," *New York Times*, 6 August 2001.

6. The joint editorial stating the new policy can be found in several journals (for instance, DeAngelis et al. 2004). It is also available online at: http://jama.ama-assn.org/cgi/content/full/292/11/1363.

7. See Dickersin and Rennie (2003) for a discussion.

8. It should be noted that Jerome Kassirer, the preceding editor, sharply disagreed (McKenzie 2002).

9. For instance, Lemmens and Miller (2003) explore the use of criminal law in regulating conflict of interest.

10. Angell (2004, 75). Relevant information can be found at the FDA Web site: www.fda.gov/cder/rdmt/pstable.htm.

11. See Angell (2004, 40). The Public Citizen Web site contains relevant information on the topic: http://www.citizen.org/.

12. Sheldon Krimsky is a strong critic of current medical research and champion of extensive regulation. He remarked that "no responsible voices call for an end to corporate sponsorship" (2003, 51). I hope the reasons that I've given will make my proposal seem sensible, but it is a sign of the times, especially in the United States, that advocating a return to the pre-1980 funding situation is called "irresponsible."

REFERENCES

Angell, Marcia. 2004. *The Truth About the Drug Companies: How They Deceive Us and What to Do About It*. New York: Random House.

Barnett, Antony. 2003. "Patients Used as Drug 'Guinea Pigs.'" *Observer*, Sunday, 9 February 2003. www.observer.co.uk/politics/story/0,6903,891938,00.html).

Bloor, David. [1976] 1991. *Knowledge and Social Imagery*. 2nd ed. Chicago: University of Chicago Press.

Brown, James Robert. 2000. "Privatizing the University—The New Tragedy of the Commons." *Science* 290:1701–2.

———. 2001. *Who Rules in Science? An Opinionated Guide to the Wars.* Cambridge, MA: Harvard University Press.

———. 2002. "Funding, Objectivity and the Socialization of Medical Research." *Science and Engineering Ethics* 8:295–308.

———. 2004. "Money, Method, and Medical Research." *Episteme* 1:49–59.

Davidson, Richard. 1986. "Sources of Funding and Outcome of Clinical Trials." *Journal of General Internal Medicine* 12 (3): 155–58.

DeAngelis, Catherine, et al. 2004. "Clinical Trial Registration: A Statement From the International Committee of Medical Journal Editors." *JAMA* 292:1363–64.

Dickersin, Kay, and Drummond Rennie. 2003. "Registering Clinical Trials." *JAMA* 290:516–23.

Drazen, Jeffrey M., and Gregory Curfman. 2002. "Editorial." *New England Journal of Medicine* 346:1901–2.

Drummond, Rennie. 2004. "Trial Registration." *JAMA* 292:1359–62.

Dunn, Andrea L., et al. 2005. "Exercise Treatment for Depression: Efficacy and Dose Response." *American Journal of Preventive Medicine* 28 (1): 1–8.

Friedberg, Marc, et al. 1999. "Evaluation of Conflict of Interest in New Drugs Used in Oncology." *JAMA* 282:1453–57.

Hacking, Ian. 1995. *Rewriting the Soul.* Princeton: Princeton University Press.

———. 1998. *Mad Travellers: Reflections on the Reality of Transient Mental Illnesses.* Charlottesville: University Press of Virginia.

Harris, G., and A. Berenson. 2005. "10 Voters on Panel Backing Pain Pills Had Industry Ties." *New York Times,* 25 February 2005.

Healy, David. 2001. *The Creation of Psychopharmacology.* Cambridge, MA: Harvard University Press.

Kondro, Wayne, and Barbara Sibbald. 2004. "Drug Company Experts Advised Staff to Withhold Data About SSRI Use in Children." *Canadian Medical Association Journal* 170 (5): 783.

Krimsky, Sheldon. 2003. *Science in the Private Interest.* New York: Rowman and Littlefield.

Lemmens, Trudo. 2004. "Piercing the Veil of Corporate Secrecy About Clinical Trials." *Hastings Center Report* 34 (5): 14–18.

Lemmens, Trudo, and Paul B. Miller. 2003. "The Human Subjects Trade: Ethical and Legal Issues Surrounding Recruitment Incentives." *Journal of Law, Medicine and Ethics* 31:398–418.

McKenzie, J. 2002. "Conflict of Interest?" ABC News.com (12 June 2002).

McSherry, Corynne. 2001. *Who Owns Academic Work?* Cambridge, MA: Harvard University Press.

Merton, Robert K. 1973. "The Normative Structure of Science." In *The Sociology of Science: Theoretical and Empirical Investigations,* 267–78. Chicago: University of Chicago Press.

O'Meara, Kelly Patricia. 2001. "Misleading Medicine." http://insightmag.com/archive/200104301.shtml (accessed 1 June 2001).

Pollack, Andrew. 2005. "Medical Researcher Moves to Sever Ties to Companies." *New York Times,* 25 January 2005.

Press, Eyal, and Jennifer Washburn. 2000. "The Kept University." *Atlantic*, March 2000. http://www.theatlantic.com/issues/2000/03/press.htm.

Proctor, Robert. 1999. *The Nazi War on Cancer*. Princeton: Princeton University Press.

Rampton, Sheldon, and John Stauber. 2001. *Trust Us, We're Experts*. New York: Tarcher/Putnam.

Shulman, Seth. 1999. *Owning the Future*. New York: Houghton Mifflin.

Stelfox, Henry Thomas, et al. 1998. "Conflict of Interest in the Debate over Calcium-Channel Antagonists." *New England Journal of Medicine* 338:101–6.

Shapin, Steven. 1994. *A Social Theory of Truth*. Chicago: University of Chicago Press.

Whittington, Craig, et al. 2004. "Selective Serotonin Reuptake Inhibitors in Childhood Depression: Systematic Review of Published Versus Unpublished Data." *Lancet* 363:1341–45.

Willman, D. 2003. "Stealth Merger: Drug Companies and Government Medical Research." *Los Angeles Times*, 7 December 2003.

Woolhandler, S., T. Campbell, and D. Himmelstein. 2003. "Costs of Health Care Administration in the United States and Canada." *New England Journal of Medicine* 349 (2003): 768–75.

10

SCIENCE IN THE GRIP OF
THE ECONOMY

On the Epistemic Impact of
the Commercialization of Research

MARTIN CARRIER
Bielefeld University

THE TERM "KNOWLEDGE SOCIETY" IS FREQUENTLY taken to express that science is among the chief economic resources of the modern world. At the onset of the twenty-first century, science assumes an economic role similar to that of coal and steel in the nineteenth century. Whether or not this picture is accurate, it is an undisputable fact that science is an important factor in economic development. In this context, science is not valued because it contributes to deciphering the code of nature but rather because of its practical impact. It is not epistemic virtues that are primarily valued but rather the capacity to intervene in natural phenomena and to use this capacity for the betterment of the human condition. This means that the progress of science is not seen as an end in itself but rather as an instrument for promoting economic growth. Research expenses are regarded as an investment, which is expected to result in appropriate financial returns.

As a consequence, scientific goals are increasingly intertwined with economic interests. A large part of research in chemistry, biochemistry, or pharmacology is carried out in private companies. In such areas, a significant portion of innovative work has moved out of the universities and into industrial laboratories. We are witnessing a thoroughgoing commercialization of research, which confronts us with the question of how the dominance of economic interests affects science and society as a whole.

Prima facie, there are three elements characterizing commercialized science. First, the research agenda is set by economic goals, and such goals may differ from epistemic objectives and systematically diverge from the considered interests of society at large. Second, industrial research tends to be carried out behind closed doors. In contrast to epistemic science, commercial research and development is not committed to unrestricted access and general scrutiny. Third, research on practical questions is assessed on purely pragmatic grounds. The sole criterion of success is that some device operates reliably and efficiently; no further epistemic ambitions are pursued. In sum, this commercialization process can be expected to lead to a biased research agenda, keep public science out of corporate laboratories, and induce methodological sloppiness. I wish to explore to what extent such worries are justified and how adverse effects on science can be kept under control.

Shifts in the Research Agenda

The emphasis on application entails that the research agenda is largely shaped by factors like practical promise, technical feasibility, and probable profit. Epistemic research projects receive, as a rule, much less attention—and funding—than applied ones. Take high-temperature superconductivity as an example. This effect turned out to be inexplicable by the so-called BCS-model that accounts for usual, low-temperature superconductivity. Research activities in this field only rarely address the challenge of giving a theoretical explanation of high-temperature superconductivity. Rather, most of the relevant efforts are directed at practically important issues such as increasing the temperature threshold at which superconductivity sets in, finding suitable substances or improving their technically relevant properties. Judging from such cases, it is undeniable that practical challenges are heading the research agenda of science.

Applied research is performed in considerable measure in industrial companies and is carried out and paid for in order to earn financial returns. It is not the quest for knowledge that drives such endeavors ahead; rather, research is considered an investment whose returns are expected to cover the costs (Rosenberg 1990, 165; Rosenberg 1991, 345; Carrier 2002). It is true that the flow of private capital and the economic aspirations that motivate the underwriting of research do much to stimulate technological progress, and even the advancement of science. But the obvious disadvantage is that the agenda of applied research is shaped by commercial interests whose pursuit is unlikely to benefit everyone concerned. The clearest examples can be found in medical research, where the bias of the agenda is undeniable and obvious. Ailments troubling the population of the rich countries top the list of challenges; third-world illnesses don't play a significant role. Philip Kitcher speaks of the "10/90 gap," which means that only 10 percent of biomedical research resources are devoted to diseases that cause 90 percent of the suffering worldwide. Curing malaria is much further down the priority list than treating obesity. In this vein, Kitcher considers it a disaster that science has entered the marketplace (Kitcher 2002, 570).

The same argument applies to economics. James R. Brown provides an example to the effect that powerful financial institutions can afford to sponsor research on the influences of high taxation on productivity. Assume the truthful result was that an increased level of taxation tends to reduce productivity, and that nothing but this truth was stated in the research report. The failure still is that other effects of high taxation are left out of consideration. Since no well-funded foundation called "The Consortium of Single Mothers on Welfare" underwrites research projects examining the issue of taxation from a different angle, the prospects of increased state income for developing poor families remain unexplored. Although nothing untrue was said, the preferred treatment of the productivity issue tends to eclipse certain contravening interests in society (Brown 2000, 1701; Brown 2001, 210).

Commercial Research and Development behind Closed Doors

A second adverse effect of the commercialization of science is a tendency to keep research outcomes secret. As Nathan Rosenberg put it, the serious possibility emerges "that the potentially great commercial value of scientific findings will lead to a loss of free and frank communication among university faculty, and a reluctance to disclose research findings from which other faculty or students might derive great benefit. Such developments could prove to be harmful to future progress in the realms of both science and technology, as well as to education itself" (Rosenberg 1991, 340). Robert Merton claimed "communalism" (as it is now often called, or "communism," as Merton put it) to be one of the pivots of the cultural value system that characterizes scientific inquiry. "Communalism" means that scientific knowledge is in public possession; knowledge cannot be owned by a scientist, nor by his or her employer. It is an essential and indispensable part of the "ethos of science" that scientific findings are public property. Merton demanded, consequently, that scientific knowledge should be accessible to everyone and suggested that the system of patenting is in conflict with this value (Merton 1942, 273–75).

The commitment to openness of research and general access to the knowledge gained is constitutive of epistemic science. The reward system of epistemic science is based on publication. By contrast, industrial research and development projects are intended to produce knowledge that can be used exclusively. After all, companies are not eager to pay for gaining knowledge that can be used afterward for free by a competitor (Dasgupta and David 1994, 495–98). As a result, important domains of scientific activity are staked out or constrained by industrial secrets or patents. Not infrequently, scientists are prohibited by their working contracts from disclosing discoveries they have made. The commitment to secrecy governs parts of industrial research; it tends to reinstate the traditional separation between scholarship and society at large that was gradually resolved in the course of the scientific revolution. The commercialization of research may go along with a privatization of science that compromises the public accessibility of knowledge (Concar 2002, 15; Gibson, Baylis, and Lewis 2002). By operating behind a veil of secrecy, commercialized research tends to reinstate an institutional form of science that is attributed to Pythagoras, its organization, namely, as brotherhoods of erudition. In this Pythagorean tradition, the truth is considered too lofty to be exposed to the stupidity of the hoi polloi. Rather, the insights gained by the sages are not for general dissemination; they are to be handed over only to the true and sincere disciples. The Pythagorean understanding of science is shared by Copernicus and Newton and so extends well into the modern era. The fruits of knowledge are reserved for the intellectual elite; they are spoiled by vulgarization.

It goes without saying that the new conspiracy of silence springs from different motives. It has less to do with being devoted to erudition and more with being attracted by worldly goods. But what is at stake here belongs to the essentials of scientific method, as it is understood after the demise of the Pythagorean ideal. Knowledge claims in science should be subject to everyone's judgment. The intersubjective nature of scientific method demands public tests and confirmation. Social epistemology has highlighted the fact that knowledge is produced by the scientific community as a whole. Science requires publicly recognized venues in which knowledge claims are subjected to criticism and control (Longino 2002, 129, 153). Consequently, secrecy may block putting knowledge claims to severe scrutiny. Such claims are neither tested as critically as they would be if the claims or the pertinent evidence were more widely known, nor can the new information be employed in related research projects. Secrecy thus tends to impede the epistemic authority and the progress of science.

The sequestration of research outcomes is not only epistemologically problematic, it is also ethically questionable. Innovations of industrial research are based on and made possible by the prior work of many other scientists. Today's findings depend on earlier discoveries, most of which were made within the public domain. Thus, withholding useful discoveries from the scientific community and from society at large is refusing to compensate for earlier benefit; it is refusing to do unto others as you would have them do unto you. Science encompasses an integrated web of intertwined doctrines and skills. This web cannot be divided neatly into pieces of information that could be owned separately. The essential participation of the scientific community as a whole—past and present, public and private—in the growth of knowledge gives us the right to know.

Tentative Epistemic Strategies and the Dominance of Practical Goals

Applied research is driven by technological interests and directed toward practical goals. As a result, applied research can be expected to be governed by a purely pragmatic attitude according to which the proper functioning of a device is the sole measure of success. If a gadget works, everything is fine; no further questions will be asked. Epistemic challenges that transcend immediate practical needs are likely to be ignored. As a result of this purely pragmatic attitude, applied research appears methodologically deficient as it heavily restricts the scope of theorizing and explaining. Not infrequently, for instance, applied research is content with contextualized causal relations. Such relations are confined to "normal" conditions and fail to address the underlying causal processes. Large portions of research outcomes on the effects of medical drugs are of this sort. Such research reports state that certain substances have particular toxic or curative effects without taking the pertinent causal processes into account. The results derive from schematic screening procedures and are not based on a theoretical understanding of the molecular mechanisms at hand. Contextualized relations of this sort are usually sufficient for bringing about the desired effect; no integration into a wider theoretical framework nor the elucidation of the relevant causal steps is sought.

The chief objective of applied research is practical success, that is, the control of natural phenomena and intervention in the course of nature. The dominance of this pragmatic attitude plausibly translates into the prevalence of tentative epistemic strategies that tend to cut off research from any deeper epistemic aspirations. Applied research expectedly is methodologically deficient (Carrier 2004a, 2004b). In the following, I wish to explore whether applied research is indeed characterized by biased agenda-setting, secrecy, and methodological sloppiness, and, if it is, to inquire into the consequences and examine possible remedies.

Commercialized Science and Normative Conceptions of Knowledge

I argued before that the commercialization of science entails that the research agenda is determined by markedly nonuniversal purposes and interests. Companies and corporations draw up the list of research topics and thus determine what is worth being studied. It follows that as regards application interests, commercialized science fails to involve a fair representation of the needs and goals of all people affected by the research.

Kitcher demands that research be guided by the ideal of "well-ordered science." In well-ordered science, the research agenda is established on the basis of the preferences of the citizens of a society. Kitcher envisages a process of deliberation in which legitimate research topics are determined by mediating the interests of the members of a society (Kitcher 2001, 117–23). Kitcher stresses that the ideal of well-ordered science suggests resistance to the dominance of research by economic companies. In commercialized science, the preferences of only a few members of a society are brought to bear on singling out suitable topics. Research pursued under the pressure of market forces is likely to be shaped by the nonmediated interests of "the rich and powerful." Kitcher's conclusion is that science ought to keep away from the marketplace (Kitcher 2002, 570). Thus, he objects to two mentioned features of commercialized science, namely, biased agenda-setting and restricted public access.

Helen Longino likewise stresses the commitment of science to democratic inclusion. But unlike Kitcher, she is not chiefly concerned with research topics but with the appraisal of hypotheses and theories. Epistemically acceptable knowledge claims need to pass a process of critical scrutiny in which objections from all relevant perspectives are taken into account. In order to make sure that the criticism leveled against an assumption is taken from the broadest possible range, all members of the society need to be incorporated in the process of evaluation—albeit not necessarily to the same degree. Knowledge grows out of a process of critical discussion that is all-embracing in that nobody willing and able to participate is excluded (Longino 2002, 129–34).

Brown advocates a similar social model of scientific scrutiny. As he argues, we all hold unwarranted background assumptions, and we are unaware of most of them. For this reason, scientific hypotheses should be tested against a wide and varied range of positions. Since we are not able to get rid of all such unwarranted suppositions, different points of view should be brought to bear on testing scientific claims. This can be ascertained, among other things, by seeing to it that scientists with potentially competing interests perform tests of a hypothesis. Pluralism in and democratization of science are necessary for epistemic reasons, in that they contribute to improving the reliability of scientific results (Brown 2001, 187).

The pluralist understanding of scientific objectivity and rationality, as advocated by Longino and Brown, stands in marked contrast to a more traditional approach going back to Francis Bacon. The Baconian model conceives of objectivity and rationality as adequacy to the situation or the facts at hand. No factors intrude into the consideration which do not represent factual elements of the problem situation. This Baconian model requires us to eliminate unfounded prejudices, whereas the pluralist model directs us to control prejudices. The pluralist model—going back, chiefly, to Popper and Lakatos—considers it an epistemic virtue to take conflicting views into account. Different biases are supposed to keep each other in check.

Longino claims to subject science to a more thoroughgoing democratization than does Kitcher. In her view, Kitcher limits democratization to agenda-setting, whereas she addresses the epistemic practices themselves. In spite of such possible differences, all these conceptions of objective or well-ordered science agree that two of the mentioned features of commercialized science are unacceptable. Science done in private and behind closed doors invites neither public agenda-setting nor general participation in the assessment of epistemic credentials. Commercialized science contravenes these normative conceptions and comes out as morally and epistemically deficient.

Given that restrictions in agenda-setting and public access are unwelcome features of science, the hard question is what can be done about it. The democratization of science would no doubt provide an effective antidote. But organizing science in such a way that it takes into account everyone's objections and considered interests would be a long-term endeavor. Therefore, it might be worthwhile to examine whether there are processes within applied science that tend to constrain the adverse impact of economic influences and keep science in accordance with such normative conceptions. In what follows I wish to explore mechanisms inherent to science that are apt to contribute to keeping commercialized research objective, public, and epistemically respectable.

Plurality, Reliability, and the Epistemic Integrity of Applied Research

An insight shared by Karl Marx and Milton Friedman is that economic forces are powerful and hard to overcome. Confronting these forces head-on may turn out to be a vain attempt. Commercialization is like a tidal wave: it can hardly be stopped, but it can be channeled into a distinct direction. Over time this might make a noticeable difference. The odds of achieving such a reorientation of science increase if one takes advantage of procedures and mechanisms that are part of applied research anyway. This can be done by exploring how far-reaching and grave the problems with biased agenda-setting, general accessibility, and methodological decline really are and whether there are factors within commercialized science that could limit their import.

To begin with, it is worth noting that the commitment of industrial companies to science leads to increased research and contributes to scientific progress. Large amounts of research in physics, chemistry, and biology would have never been undertaken if no economic interests were at stake. In some fields of research, the alternative to commercialized science is not research directed at the common good, but rather no research at all. Presumably, in some such cases of undone research, the ensuing lack of knowledge would be detrimental to public welfare. The production of knowledge is an important asset and merits appreciation. I grant that the dominance of economic interests on research in industrial enterprises has unwelcome side effects, such as limited access to results. Still, emphasis should not be placed exclusively on such shortcomings in the distribution of knowledge. It should be recognized in addition that some knowledge would otherwise not have been produced in the first place.

A pivotal issue is whether research conducted by private companies and driven by practical challenges retains its epistemic integrity. The objection from the pluralist conception of rationality says that industrial research tends to lack external controls and is thus prone to come up with less reliable results than research done in public. Another methodological worry to the same effect is that industrial research might suffer from its supposedly purely pragmatic attitude. The heavy pressure to come up speedily with working solutions could promote reliance on "quick and dirty" procedures and thus contribute to reducing standards of epistemic quality control.

There is reason to believe that epistemic trustworthiness and respectability do not suffer dramatically from the dominance of research driven by economic interests. First, in view of the pluralist model of scientific objectivity and rationality, it is not necessary—nor even possible—to relinquish one's prejudices. Scientific objectivity need not be based on personal virtues of the scientists; in particular, it does not require their "disinterestedness" (as the methodological canon of the seventeenth and eighteenth centuries assumed). What matters, instead, is control of judgments and interests by bringing to bear contrasting judgments and interests. Competition among companies could contribute to stimulating such reciprocal control. As a rule, a number of corporations competitively struggle for a given innovation. Competition serves to provide a type of pluralism resembling the sort approved by the pluralist conception of rationality. For instance, if research proceeds in a false direction in a company, there is a good chance that the involved scientists may be proven wrong by the more successful research product of a more successful competing company. Second, reliability is of paramount concern for applied science of all sorts. There is a strong economic pressure on industrial research to yield trustworthy results. If the gadget fails, you may go broke. Industrial research cannot afford to be unreliable. However, contextualized causal relations are less reliable than theoretically understood generalizations. The reason is that contextualized relations are confined to normal circumstances, and it frequently happens that the presence of distorting factors invalidate such relations and make them unsuitable as a basis for intervention. The antibiotic efficacy of penicillin is a contextualized causal relation that was vitiated by the emergence of resistance. Bringing such disturbing influences under control requires understanding of the causal chains at hand. The practical objectives of gaining or maintaining the power of intervention drives applied research toward epistemic goals like causal understanding.

Therefore, a lack of objectivity or reliability is not a pressing problem of applied research. I admit that there are cases in which the interests of a sponsor of a study have biased its outcome (Brown 2002, 297–98). If you choose to consult research reports issued by the tobacco industry on the risks of smoking, you can hardly expect to obtain a balanced and complete picture. But the two considerations outlined contribute to alleviating the suspicion of deceptiveness. First, whereas studies such as the one mentioned are mostly for public effect, the case is different with research intended to result in new marketable goods. Reliability of outcome is among the preconditions for the technological use of research. Second, even regarding studies whose chief intention is to affect public opinion, competition among companies serves to expose faulty results. If one company publishes research reports that give a false lead to their own product, the competitor will be glad to disclose the mistake. As a result, the forger would suffer from a loss of credibility. That is, competition in the economy is a factor that might tip the scales in favor of truth. In conformity with the pluralist notion of objectivity, competing interests serve to reciprocally correct the shortcomings of interest-guided research, and reliability is achieved by bringing contrasting voices to bear on a subject. The good news is that pluralism is to a certain degree realized in commercialized research, so that pluralism can be accomplished prior to a full-scale democratization of science.

I admit at once that this strategy has its limits. In a number of cases, pluralism is conspicuous by its absence, as the long-standing, unanimous denial of risks involved in smoking by the tobacco industry amply demonstrated. Moreover, there is one area where a strong public grip on the pertinent research is mandatory, namely security issues. The reason is that in this area the possible damage is unacceptably high and the risks involved far outweigh any gains due to the stimulation of progress by competition and economic interests. For instance, studies on the risks involved in chemical substances must not be carried out by companies eager to manufacture these substances. There is reason to be concerned that business interests might influence judgments about the appropriateness of results (Machamer and Douglas 1999, 49). It is not necessary to assume outright fabrication of data to be worried about leaving it to private companies to provide risk-related evidence. Risks are statistical in nature, and judgments about statistical tendencies are affected by a number of circumstantial factors. For instance, if a sufficient number of studies is performed, you can trust that one of them will yield statistically significant results, or insignificant ones, as the case may be. It follows that you need not falsify data in order to create misleading conclusions, you only need to single out studies or withhold information.

For this reason, it is imperative that security-relevant decisions are based on independent, complete, firsthand access to the pertinent effects and circumstances. In addition, it needs to be ascertained that security issues are dealt with by institutions that give the interests of potential victims appropriate weight.

Both considerations suggest that public authorities should take responsibility for matters of safety and that they are granted the right to establish on their own the evidence to assess risks for human health or environmental conditions. In matters of security, a zero-failure policy is mandatory, and this demands tight public control.

On the Difficulty of Restricting an Agenda to Goals Fixed in Advance

Another pertinent observation is that it is difficult to keep a research agenda restricted to goals and questions that were fixed in advance. Rather, research often develops a question dynamics that spawns unforeseen issues and results. A scientific project is often characterized by uncertainty; you never know for sure what will come out of it. Industrial project managers are no more capable of constraining the issues addressed to a small area fixed beforehand than university researchers are. Research is creative and cannot be fenced in easily. Thus, sticking to a given research agenda is harder to achieve than it might appear at first sight.

The problem-driven shift in the research agenda is underscored by a feature I call application innovation. It involves the emergence of theoretically significant novelties within the framework of use-oriented research projects (Carrier 2004a). Not infrequently, practical challenges raise fundamental issues in their train. The challenges cannot appropriately be met without grappling with the issues. Although theoretical understanding is not among the objectives of applied research, it may yet be produced in the course of solving practical problems. On some occasions, treating such problems appropriately demands addressing epistemically significant issues. Once in a while, applied research naturally grows into basic science and cannot help generating epistemic insights.

For example, high-temperature superconductivity was discovered in 1986 in the IBM research laboratory near Zurich, and its identification stimulated the development of new theoretical accounts of superconductivity, if at a moderate level of intensity. Similarly, the transistor effect was found in the Bell laboratories. The emergence of this effect was based on the truly innovative procedure of adding impurities to semiconductors, which act as electron donors or acceptors. This idea enriched solid-state physics tremendously. To turn to biology, the path-breaking polymerase chain reaction was first conceived in a biotechnology firm, and the revolutionary conception of prions was elaborated in the practical context of identifying infectious agents. Prions are infectious proteins that reproduce without the assistance of nucleic acids; they were discovered during a use-oriented study on the sheep disease scrapie.

In these examples, research had been directed toward some practical goal, but it led to a profound theoretical innovation. In fact, large parts of the innovations in basic research done in industrial companies were produced unintentionally; they arose as unplanned by-products of wrestling with practical issues. Industrial attempts to come to terms with applied challenges not infrequently bring about theoretical breakthroughs (Rosenberg 1990, 169–70).

This is no accident. Applied research tends to transcend applied questions for methodological reasons. A lack of deeper understanding of a phenomenon eventually impairs the prospects of its technological use. Superficial empirical relations, bereft of theoretical understanding, tend to collapse if additional factors intrude. Uncovering the relevant mechanisms and embedding them in a theoretical framework is of some use for ascertaining the applicability of a finding. Scientific understanding makes generalizations robust in the sense that the limits of validity can be anticipated or, as the case may be, expanded.

The upshot is that treating applied questions appropriately requires that they are not treated exclusively as applied questions. The attempt to gain reliable, practically relevant knowledge evolves into a dynamical process that tends to lead into fundamental issues. Consequently, applied research, as a rule, honors the epistemic respectability of science as a knowledge-seeking enterprise. At the same time, the problem shifts involved in innovative applied research demonstrate how difficult it is to keep a research agenda within a fixed topical realm. Creative research cannot be planned, and this is why constraining the framework of questions, issues, and challenges eventually is beyond the forces of industrial companies.

On the Difficulty of Keeping Research Outcomes Classified

Another feature of commercialized research that runs counter to the traditional notion or ethos of science concerns restrictions in the access to scientific discoveries. Privately sponsored research tends to operate behind closed laboratory doors and thereby compromises the commitment to universal availability that has been characteristic of the value system of science ever since the scientific revolution. The intention to keep research outcomes classified and to confine the use of knowledge is often regarded as the distinctive feature of industrial research. The reward system of academic science is based on the public recognition of discoveries, which in turn presupposes that the insights gained are disclosed and made generally available. In contrast, the hallmark of success of research and development in the technological sector is financial benefit accruing from putting some piece of information to practical use. Knowledge gain is not an end but an instrument designed for exclusive use (Dasgupta and David 1994, 495, 499–501).

Commercial research and development is often governed by the obligation to keep silent. From the epistemic point of view, such restrictions in the availability of knowledge constitute a worrisome feature. However, there are counteracting mechanisms that tend to push commercial research toward openness. First, it is hard to fence in ideas. In contrast to most material commodities, knowledge is a nonexclusive good. It can be possessed by anyone who takes the pains to acquire it. Knowledge is expansive in character, keeping it within the confines of a company is difficult to accomplish. Second, progress is facilitated by cooperation. If two researchers each solved half of a given problem, sharing their knowledge may make them realize that they are done, whereas a lot of work would still be left to do if each one had proceeded separately. Conversely, pursuing similar research endeavors in isolation from each other may have the unwelcome effect of making the same discovery twice. Sequestration may be a costly impediment to commercially successful research whereas cooperation may pay off economically. Cooperation can be realized either by large research teams or by publishing the results (Wray 2002, 155–58). This means that sharing thoughts tends to benefit all parties involved, which supports a policy of open labs.

Indeed, some features of present-day commercial research and development practice bear witness to the recognition that keeping research outcomes classified could hurt a company. Let me mention three incentives for going public: the exploitation of the academic sector, the protection of intellectual property rights, and the mass media attention to science.

First, perverse as it may sound, the exploitation of the academic research and education system by commercialized science serves to promote public access to its findings. One relevant effect is that taking advantage of the results produced by publicly funded fundamental research requires deeper understanding. After all, scientists in private companies are expected to assess the implications of this research for possible technological innovations. Such an advanced understanding requires being locked into the pertinent research network. The reason is that only part of the relevant knowledge is codified explicitly. Part of the know-how is tacit knowledge. It is difficult to build a working device only on the knowledge laid down in the patents that purportedly describe the operation of that device. The skills necessary for taking advantage of discoveries made elsewhere are acquired most conveniently by being another player in the field, that is, by doing fundamental research oneself. In sum, public fundamental research is exploited most efficiently by private institutions if the latter also pursue fundamental research projects. But this demands making one's discoveries generally accessible. In fact, a lot of private laboratories do publish their findings and seek recognition as scientifically reputable institutions. As a result, a reciprocal exchange between academic and industrial research can be observed (Rosenberg 1990, 171; Dasgupta and David 1994, 494; Nichols and Skooglund 1998).

Private aspirations to take advantage of the public education system work in the same direction and likewise contribute to a policy of openness. In Germany it frequently happens that a company underwrites the work for bachelor's, master's, or Ph.D. theses. In exchange, they are granted the privileged use of the results. But this exclusive use expires after a limited time period, and the findings are published. After all, they are supposed to be the basis for conferring academic degrees and need to be publicly accessible for this reason.

I do not deny that it would be better if the knowledge were generally accessible from the outset. On the other hand, it would be worse if the knowledge had never been produced. Given severe public budget restrictions, private funding contributes to securing academic research. Industrial sponsoring makes certain epistemic research projects feasible in the first place.

Second, patenting procedures are contentious and have the reputation of depriving science of its public nature. Patents grant exclusive use of an invention and therefore contradict Merton's "communalist" ethos of science. In order to appreciate the import of this apprehension, three items deserve to be noted: in Europe only technical procedures can be patented, but not their outcome; patenting involves disclosing the invention, and the protection granted expires after a certain period (usually twenty years); without the economic stimulus provided by patent protection, the pertinent research would probably not have been carried out. Without patent protection, companies doing no research would be able to take a free ride and reap the harvest while saving the cost of the seed. Patenting provides an economic incentive for private research and contributes to making scientific information more widely available. Without patent protection, the only means for securing exclusive use would be secrecy.

Paradoxical as it may sound, granting intellectual property rights contributes to making research generally accessible.

It is true that the refusal to publish before a patent application has been filed involves withholding information, at least temporarily. But the financial benefit that patenting procedures grant to successful research stimulates scientific progress (Munthe and Welin 1996, 417). Furthermore, the obligation to disclose the patented innovation to the public advances the openness of science. Patenting does not protect the information itself but only its commercial use.

Patenting promotes general accessibility by another strategy called "defensive publication." If a company publishes some finding, it becomes part of the state of the art against which a competitor's patent application is judged. A defensive publication obstructs such an application by rendering the invention lacking in novelty. Defensive publication is used as a protection against a new product of an unanticipated kind introduced by a competitor. Defensive publication serves to narrow the range of procedures that a competitor can possibly have protected by a patent (Barrett 2002, 191; see also Rosenberg 1990, 171). Although it appears odd at first sight, granting exclusive use of information may promote its general availability.[1]

Third, scientists from private and public research institutions alike seek the attention of the general public; they are keen on being in the media. It is increasingly part of the business to be widely known. In contrast to the assumed tendency to secretiveness, scientists rush to obtain notice in the mass media. One of the side effects of media attention is the risk of compromising scientific credibility (Weingart 2001, 244–53). Insufficiently confirmed ideas are aired publicly and presented prematurely as scientific knowledge. But with respect to the issue of secrecy, the effects of media exposure are clearly to be welcomed. The run toward public recognition is quite the reverse of excessive secrecy.

All three features of latter-day applied science suggest that it is more difficult than one might have expected to keep one's research outcomes hidden from the public eye.

Conclusion

I began by listing three potentially harmful features of science in the grip of economic forces: a biased research agenda, secrecy over research outcomes, and a neglect of epistemic aspirations. First, the problems worth being addressed are selected according to the economic prospects they offer. Second, commercialized research tends to operate behind a veil of secrecy. Instead of the commitment to openness and general scrutiny that characterizes epistemic science, industrial research is directed at exclusive use of scientific innovations. Third, it is feared that commercialized science is governed by a pragmatic attitude which renders science superficial and makes it abandon in-depth studies.

The discussion has shown that there is reason for concern about the state of commercialized science, commissioned research, or research conducted in private companies for economic reasons alone. All of these worries are justified, indeed, but to a different extent and, on the whole, to a lesser degree than was initially feared. The thrust of my argument is that there are features in applied research that reduce the impact of commercial forces and are suitable as safeguards against some of these adverse tendencies. First, keeping a research agenda within boundaries which were fixed in advance is not easy. Research is often distinguished by a question dynamics that drives a research agenda beyond anticipated topical limits. Second, it turns out that the attempt to keep research outcomes secret backfires and damages those who try. Laboratories operating behind closed doors are cut off from the benefits of cooperation. Observation confirms this pressure toward sharing knowledge in applied research. The amount of classified outcome is smaller than thought beforehand. Third, methodological deficiencies are likewise less pronounced than assumed in advance. The question dynamics built into successful practical research drives science toward epistemic challenges.

It is to be acknowledged that market forces in themselves fail to make sure that research outcomes are epistemically trustworthy and suitable for promoting the common good. I do not suggest that the free market can take care of epistemic and social demands. However, it is also true that some internal processes in applied science run counter to the adverse tendencies of commercialization. These processes reduce the impact of economic factors on science and can be employed to further limit their undesirable consequences. Regulating market forces might be helpful for keeping epistemically and (to a smaller degree) socially harmful influences at bay while allowing us at the same time to take advantage of the advanced level of creativity and the increase of novelty that tend to go along with competition. In the end, there is some sort of connection, if fragile, between becoming wise and getting rich or between the institutional goals of deepening our knowledge and increasing our wealth. The mitigating processes from within commercialized research notwithstanding, it remains true that we need a publicly sponsored science that is devoted to openness and the legitimate interests of everyone concerned. This is an continuing challenge for university research.

NOTES

I am grateful to Matthias Adam for his valuable suggestions, which did a lot to improve the argument.

1. It is true that patents are sometimes employed for blocking technological progress rather than advancing it. But at least this lamentable strategy, which was criticized already by Merton (1942, 275), is constrained by the rules of patenting. After the protection period has expired, the invention can be used freely by everybody, and the potential benefit of being the first in a technological field would be gone. But then it might be better to take advantage of the head start granted by the patent. Moreover, patenting is not confined to private companies; public institutions likewise seek patents. A number of state-run research institutes in Germany are required to earn their budget by patents and commissioned research. Likewise, the Atlanta Center for Disease Control presently seeks a patent on the identification of the RNA sequence of the SARS virus, and it is said to be developing a marketable SARS test. As state budgets are in dire straits, the financial pressure on public research institutes makes them act like private companies.

REFERENCES

Barrett, Bill. 2002. "Defensive Use of Publications in an Intellectual Property Strategy." *Nature Biotechnology* 20:191–93.

Brown, James R. 2000. "Privatizing the University—The New Tragedy of the Commons." *Science* 290:1701–2.

———. 2001. *Who Rules in Science?* Cambridge, MA: Harvard University Press.

———. 2002. "Funding, Objectivity and the Socialization of Medical Research." *Science and Engineering Ethics* 8:295–308.

Carrier, Martin. 2002."Grundlagenforschung ist kein Profit-Center." *Physik Journal* 1 (4): 3.

———. 2004a. "Knowledge and Control: On the Bearing of Epistemic Values in Applied Science." In *Science, Values and Objectivity*, ed. Peter Machamer and Gereon Wolters, 275–93. Pittsburgh: University of Pittsburgh Press / Konstanz: Universitätsverlag.

———. 2004b. "Knowledge Gain and Practical Use: Models in Pure and Applied Research." In *Laws and Models in Science*, ed. Donald Gillies, 1–17. London: King's College Publication.

Concar, David. 2002. "Corporate Science v the Right to Know." *New Scientist*, 16 March 2002, 14–16.

Dasgupta, Partha, and Paul A. David. 1994 "Toward a New Economics of Science." *Research Policy* 23:487–521.

Gibson, Elaine, Françoise Baylis, and Steven Lewis. 2002. "Dances with the Pharmaceutical Industry." *Canadian Medical Association Journal* 166:448–50.

Kitcher, Philip. 2001. *Science, Truth, and Democracy*. Oxford: Oxford University Press.

———. 2002. "Reply to Helen Longino." *Philosophy of Science* 69:569–72.

Longino, Helen E. 2002. *The Fate of Knowledge*. Princeton: Princeton University Press.

Machamer, Peter, and Heather Douglas. 1999. "Cognitive and Social Values." *Science & Education* 8:45–54.

Merton, Robert K. 1942. "The Normative Structure of Science." In *The Sociology of Science: Theoretical and Empirical Investigations*, 267–78. Chicago: University of Chicago Press, 1973.

Munthe, Christian, and Stellan Welin. 1996. "The Morality of Scientific Openness." *Science and Engineering Ethics* 2:411–28.

Nichols, Steven P., and Carl M. Skooglund. 1998. "Friend or Foe: A Brief Examination of the Ethics of Corporate Sponsored Research at Universities." *Science and Engineering Ethics* 4:385–90.

Rosenberg, Nathan. 1990. "Why Do Firms Do Basic Research (with Their Own Money)?" *Research Policy* 19:165–74.

———. 1991. "Critical Issues in Science Policy Research." *Science and Public Policy* 18:335–46.

Weingart, Peter. 2001. *Die Stunde der Wahrheit. Zum Verhältnis der Wissenschaft zu Politik, Wirtschaft und Medien in der Wissensgesellschaft*. Weilerswist: Velbrück Wissenschaft.

Wray, K. Brad. 2002. "The Epistemic Significance of Collaborative Research." *Philosophy of Science* 69:150–68.

11

PROMOTING DISINTERESTEDNESS OR MAKING USE OF BIAS?

Interests and Moral Obligation in Commercialized Research

MATTHIAS ADAM
Bielefeld University

THERE CAN BE LITTLE DOUBT THAT THE ONGOING and already well-advanced commercialization of considerable areas of scientific research is responsible for profound changes not only in the institutional but also the normative constitution of science. In particular, traditional norms such as disinterestedness or impartiality and the openness of scientific research have come under pressure. Many voices warn against the consequences of commercialization for the reliability, trustworthiness, and ultimately, progress of scientific research (for example, Ziman 2003). At the same time, other authors welcome these changes or some of their aspects, such as an increased orientation of science toward the needs of the general public or the entrepreneurial spirit taking hold of universities, and see in them the signs of a promising new mode of research (Gibbons et al. 1994; Etzkowitz 1997). The status of disinterestedness is particularly ambivalent. On the one hand, disinterestedness, or the impartiality of scientific research, appears to be an important precondition of scientific objectivity. On the other hand, the consideration of partial interests by scientific research can serve important societal needs, and self-interestedness on the part of researchers might help to advance science as a whole.

The aim of this paper is to investigate more closely what position the norm of disinterestedness actually occupies and legitimately aspires to. The study of pharmaceutical research and development is well-suited for this purpose. Many violations of disinterestedness in pharmaceutical research have been documented that have led to inappropriate, biased, or even falsified scientific results. For this reason, closer control or restraint of commercial interests has repeatedly been called for. At the same time, many legitimate interests are present in pharmaceutical research, and a number of proposed measures try to make productive use of the existing interests rather than to restrain them. I will discuss, in this chapter, both problematic violations of disinterestedness and countermeasures that have been proposed or implemented, and evaluate their bearing on the validity of disinterestedness as a moral and epistemic norm. Before turning to pharmaceutical research, I will briefly recall some of the general reasons that speak in favor of disinterestedness.

Disinterestedness as Moral and Epistemic Norm

Disinterestedness and the closely related duty of impartiality belong to those research norms which, like the obligation to publish research results, can be regarded as both moral and epistemic standards. The now classical view of the relationship of epistemic function and ethical status was formulated by Robert K. Merton ([1942] 1973). According to Merton, disinterestedness is part of the ethos of science since it serves the institutional goal of science, namely "the extension of certified knowledge" (270). Merton discusses violations of disinterestedness that aim to promote the researcher's career or particular political agendas. It is part of the scientific ethos to ban such egoistic or political interests from influencing the interpretation of data or the derivation of consequences from scientific findings, since data are otherwise likely to be falsified and conclusions drawn that are untenable by scientific standards. However, Merton is not explicitly concerned with the question whether individual scientists are morally obliged to be disinterested. Though he allows for the possibility that researchers internalize the norm and are therefore psychologically compelled to act in accordance with it, he takes the institutionalized mutual control of scientists to be the main instrument of science to secure disinterestedness and to be the main indicator of its position within the scientific ethos. According to Merton, the use of illicit means is rare in science not because scientists are particularly virtuous, but because they check and criticize each other's results and conclusions so effectively.

Still, the institutional implementation of the norm of disinterestedness by a system of mutual control is compatible with its being binding for individual scientists. By excluding the influence of distorting factors on scientific research, disinterestedness can serve scientific objectivity. It can be argued, however, that objectivity is a general moral standard of science. It can be derived not only from the general obligation to be truthful and to refrain from deception that applies to everybody, but can also be based on the particular position that science has within society. Scientists possess knowledge and capacities that allow them alone to deal with problems and topics that are of great importance to everybody. Prima facie, scientists have the duty to employ these capacities and to apply the knowledge for the good of society, which includes the provision of reliable and trustworthy knowledge (Shrader-Frechette 1994, 24, 49).

Of course, the obligation to serve the general good can also derive from the public funding of science, particularly if no specific conditions are tied to the funding. Aside from the source of funding, however, the duty to promote the general welfare is enforced if science constitutes a profession. Professions enjoy privileges like increased autonomy and far-reaching self-regulation. In return for these privileges, members of the professions are obliged to organize themselves in such a way that the services of general value that they alone can provide are secured (Bayles 1989, chap. 1). Science possesses many of the features of a profession. For instance, scientists decide autonomously about membership in the scientific community and about methodological standards. In addition, the allocation of research funds is typically based on scientific evaluations of promising research areas and peer review of research proposals. Accordingly, researchers seem to be strongly bound to observe the general good rather than partial interests and to guarantee the objectivity and reliability of research results (Shrader-Frechette 1994, 24; Resnik 1998, chap. 3). Disinterestedness is thus both an epistemological and ethical imperative.

Disinterestedness and closely connected norms such as impartiality and balanced consideration of interests have been defended not only with respect to the very conduct of scientific research, but also concerning decisions about the research agenda and with respect to the use and distribution of research results (e.g., Shrader-Frechette 1994, 55). The justification of disinterestedness sketched above can in principle be extended also to these stages of the research process. Research results can lack objectivity if the research agenda excludes questions that are of great relevance to the evolving scientific understanding of an area. Accordingly, it can be demanded that an investigation into some area of practical importance must not be restricted to questions that serve certain partial interests, but has to cover the whole area that is relevant (cf. James Robert Brown's paper in this volume). Similarly, the presentation of research results must not leave out important aspects, even if otherwise only truths are stated. The communication of research results always carries with it the expectation that everything is presented which is important for the understanding of the results or for the derivation of action and which cannot be assumed to be known to the audience (cf. Grice 1989, chap. 2). To withhold important aspects or to exclude them from a scientific investigation runs counter to the obligation of objectivity. In addition, since professional ethics demands that science not only secure the reliability and trustworthiness of its results, but also contribute to the general welfare, the choice and direction of research programs and the use and distribution of their results is subject to further ethical conditions. For instance, research must not be disinterested in the sense that it uniformly neglects all interests in practical applications, but it must consider in an impartial way legitimate interests in technological applications. Similarly, the use of scientific results in technological development must take into account the interests of those that are potentially affected.[1] The given justification of disinterestedness thus does not imply that science should restrict itself to the advancement of fundamental scientific knowledge and should neglect all interests of use, but rather sees it in the role of an impartial provider of services that are of general value.

Limits to the Validity of Disinterestedness in Pharmaceutical Research

The reasons that can be adduced in favor of disinterestedness at the same time point toward a number of limitations of the norm's validity. For instance, it will usually not be possible to demand from individual scientists or from single research groups that they alone provide a comprehensive and balanced treatment of an area of research. It might just be too much work to do. Rather, it will in many cases be the duty of a scientific discipline or of the scientific community as a whole to provide such a treatment. From this communal task, it might still follow that an individual scientist has the obligation to contribute proportionally to the comprehensive and balanced treatment, for example by addressing within the confines of her means questions that have so far been neglected. Instead, however, one could also imagine an alternative arrangement that largely frees individual scientists from choosing research topics that do not suit their own or their sponsors' interests. A reward system that places a particular premium on the investigation of neglected research topics might be capable of securing a balanced and comprehensive scientific research program while allowing scientists or research groups to follow their own preferences.[2]

Scientists might still have the duty to help establish such a reward system. Yet, if such a system is in place, the moral evaluation of self-directed interests differs considerably from the evaluation that comes from a Mertonian system of mutual control. While the system of control aims at restraining the influence of partial interests on research and is thus compatible with the moral and epistemic condemnation of the interests, the reward system makes productive use of those interests in order to bring about overall impartiality. To the extent that a system of incentives could work, the self-interestedness of scientists or their sponsors would then be sanctioned.

As observed above, the disinterestedness or impartiality of science does not presuppose that interests of potential users of scientific results are ignored. On the contrary, it was argued that science has the obligation to observe the most legitimate interests and thus to serve urgent needs. When it comes to pharmaceutical research (or more generally to biomedical research), the interests to be considered are obviously those of the patients suffering from diseases. Science clearly has the duty to help alleviate suffering. This, however, can provide further justification to the pursuit of commercial interests in research. Plausibly, commercial successes are to a considerable degree dependent on practical utility. The economic value of a new drug crucially depends on its therapeutic usefulness. Addressing some of the most urgent medical needs can therefore both be an economically viable strategy for a pharmaceutical company and serve some of the most legitimate interests (cf. Hirsch 2002).

It is a well-known fact, however, that economic and public interests in many cases do not coincide so happily. In particular, urgent medical needs in the developing countries, but also rare diseases in the industrialized countries are in many cases not met by adequate efforts from pharmaceutical companies. The large majority of corporate research is instead concentrated on the common diseases of the developed world. According to a rough estimation of the World Health Organization, only 10 percent of health research funds go into those therapeutic areas in which 90 percent of the world's medical problems lie (Ramsay 2001). In both the United States and Europe, companies can get public support and marketing exclusivity if they develop so-called orphan drugs that are likely to be indicated only for the treatment of rare diseases. However, no similar incentives exist for corporate research on diseases of the developing world such as malaria, tuberculosis, or cholera. The development history of the anti-protozoic drug nitazoxanide illustrates some of the typical obstacles for a stronger contribution of corporate research to these therapeutic areas.

Nitazoxanide is used against diarrhea caused by the protozoals *Cryptosporidium parvum* and *Giardia lamblia*. In many developing countries, diarrhea caused by these protozoals is very common, and it is also frequent in the developed countries. *Cryptosporidium parvum* is one of the most critical pathogens where drinking water is produced from surface water and is supposed to be the most commonly identified intestinal pathogen among children aged one to five (WHO 2002, 71–72). Giardiasis is a great medical problem as well, in particular among children in the developing countries. Diarrhea caused by *Giardia lamblia* infections is often chronic and leads to substantial weight loss and failure to thrive, while *Cryptosporidium parvum* can even be lethal if patients are immunocompromised (e.g., due to AIDS) or malnourished (WHO 2002, 72–81). Nevertheless, thirty years passed from the discovery of nitazoxanide until its clinical introduction. The antiprotozoal action of nitazoxanide was found by Jean-François Rossignol and Raymond Cavier at the Paris Institut Pasteur in the early 1970s (Rossignol and Cavier 1975). Despite the fact that its usefulness in pediatric diarrhea must have appeared probable from the beginning, clinical development was only really tackled after Rossignol had himself founded a pharmaceutical company in 1993. Clinical development of the drug in the United States then first aimed at a medical problem of the developed world, namely protozoal diarrhea in patients suffering from AIDS. Only after the studies foundered on insufficient patient enrollment were controlled studies conducted with children in Egypt, Zambia, and Peru. These studies demonstrated the clinical utility of nitazoxanide in pediatric diarrhea and led to the approval of the drug in 2002. Later, the indications were extended to include *Giardia lamblia* and *Cryptosporidium parvum* infections in adults (White 2003; Food and Drug Administration 2005; Romark 2006).

This case not only shows some of the difficulties of industrial development of drugs to combat diseases of the developing world, but it also supports the view that without industry engagement, most drugs would not be developed clinically.[3] Beyond the obstacles to clinical development, the long-term provision of a drug by a manufacturer can also be at risk if the market is insufficiently attractive. In some cases, the production of antiparasitic drugs by pharmaceutical companies is dependent on there being substantial markets in the developed world, while the production of other drugs has been stopped for lack of such a market (White 2003). It fits the situation that a marketing strategy of the manufacturer of nitazoxanide is apparently to argue that protozoal infections particularly among adults are underdiagnosed in the United States (Romark 2006).

While this case underpins how important it is for pharmaceutical development to be oriented toward the most urgent medical needs and not primarily toward the existing pharmaceutical markets, it remains an open question to what extent companies can be blamed for the imbalance in pharmaceutical research and development. It could certainly be demanded of industry that it address the medical needs of developing countries. Even so, companies could not be required to put their competitiveness or even existence at risk. Therefore, it would appear to be primarily a public task or an individual responsibility to provide funds for research and development on neglected diseases. As long as these funds are insufficient to meet the needs, the biased consideration by pharmaceutical companies of the interests of those suffering from conditions such as hypertension, diabetes, cancer, or Alzheimer's disease is still better, from a moral point of view, than would be a uniform neglect of all medical needs. While impartiality thus remains an obligation for pharmaceutical research and development as a whole, the pharmaceutical industry to a considerable degree seems to be exempt from this obligation and can thus legitimately pursue a partial research agenda.

These considerations indicate that there are various limits to disinterestedness as a moral standard of science, in particular when it comes to setting research and development priorities. There might be no conflict between self-directed interests and those of the general public, due to a system of incentives and an efficient market that bring self-interested research priorities in line with general welfare and urgent needs. In addition, if the research is conducted by private companies, the duty to promote the general welfare might be weakened.[4] Scientists in industry, not sharing the liberty of some of their academic peers, therefore also do not share all of their duties.

At the same time, however, the validity of disinterestedness and impartiality appears unchallenged with respect to other stages of the research process. For instance, once a company has decided in favor of a pharmaceutical research and development program, it cannot neglect investigations into the potential risks of the drugs. In addition, it does not seem to be free to let its commercial interests bias or even falsify the research, or to publish only those results that help it to market its products, or to conceal evidence of therapeutic inutility or even harmful consequences. I will now turn to some of the findings that suggest that such obligations to disinterestedness, which persist despite the considerations mentioned above, are in fact violated regularly by pharmaceutical companies.

Biasing Interests in Pharmaceutical Research

There is broad discussion on the influence of interests, in particular of a commercial kind, on pharmaceutical research and development. Much of the debate centers on clinical studies in which the efficacy and safety of drugs is investigated by human testing. Findings indicate that clinical studies are often influenced by commercial interests in an ethically unacceptable way. Investigations have shown, for instance, that new drugs have a higher probability of being evaluated positively if the study has been sponsored by industry rather than by the public or by independent organizations (Davidson 1986; Djulbegovic et al. 2000; Als-Nielsen et al. 2003). An investigation of 370 clinical studies from all areas of drug therapy shows that 16 percent of the studies sponsored by nonprofit organizations recommended the new drug as the best therapeutic option compared to 51 percent of industry-sponsored studies (Als-Nielsen et al. 2003). Similar effects of the source of funding were found with pharmacoeconomic evaluations of the cost-efficiency of therapies. Out of 20 industry-sponsored studies of new cancer drugs, only 1 (5 percent) evaluated the cost-benefit relation negatively, while from the 24 independently sponsored studies, 9 (38 percent) did. In all cases of industry funding, the study was funded by the producer of the respective drug (Friedberg et al. 1999).

There are many ways in which such one-sided results can come about. The very design of clinical studies can have great effects on the probability of positive results. For instance, promising features (such as endpoints) and particular patient populations can be chosen, even if a broader application is ultimately intended. It has also been found that in comparative studies, new drugs were given a head start by dosing standard therapies inappropriately (Montaner, O'Shaughnessy, and Schechter 2001). In addition, data collection offers many options for biased influence. To be able to enroll the desired numbers of patients, studies are often distributed over many clinical centers, while the industrial sponsor collects and analyzes the data. In such cases, clinical researchers often have no knowledge of the body of evolving data as a whole and thus cannot check whether their contributions are processed and interpreted adequately (Montaner, O'Shaughnessy, and Schechter 2001; Collier and Iheanacho 2002).

Finally, the publication of research results raises many questions. Results from clinical studies often remain unpublished. Pharmaceutical companies argue that not all results are scientifically or medically relevant and that many studies are only exploratory (Hirsch 2002). However, the supposition could also be that positive results are picked out for publication or are scheduled close to the launching of a new drug, while negative results are published with delays or are not published at all. In addition, choosing to publish study results either singly or only together with the results of other studies can influence both the overall conclusions of a paper and the attention that the study results obtain. An investigation of the publication of 42 clinical studies on five new antidepressants suggests that such fine-tuning of the publication impact does occur. As it turned out, out of the 21 studies with significant positive results, 19 studies were published alone and the other 2 studies appeared together with further results. From the 21 studies that did not show a significant effect, only 6 were published alone, 11 appeared together with further results, and 4 were not published at all (Melander et al. 2003).

Independent of the source of funding, publication is biased by the fact that negative results are less likely to be accepted for publication by journals, are more often published in journals with lower reputation, and are less likely to be cited than positive results (Collier and Iheanacho 2002).

Furthermore, the use of results from clinical studies for marketing purposes is often criticized. It has been found, for instance, that a large proportion of promotional leaflets sent to general practitioners contained medical statements that could not be checked by scientific publications or that did not adequately report the conclusions of cited studies (Kaiser et al. 2004).

Should Pharmaceutical Research Be Socialized?

These findings suggest that violations of the norm of disinterestedness are widespread in pharmaceutical research and that such violations in many cases lead to unacceptable concealment, biasing, or falsification of research results. It seems natural, therefore, to regard the corporate sponsorship and the commercial interests that come with it as a main cause for concern and to demand that pharmaceutical research should be kept free from commercial interests. James Robert Brown has made a particularly far-reaching proposal of this kind.

According to Brown, medical and pharmaceutical research should be socialized and be funded exclusively by the public. To achieve this, he proposes that corporate pharmaceutical research be deprived of its economic basis by abolishing patents on pharmaceutical innovations. This would bring with it cost reductions for drugs, and the money saved could be spent, according to Brown, on public pharmaceutical research (2002; this volume).

Brown's proposal would certainly have to be spelled out in greater detail before its impact on the various national health systems could be fully evaluated. For example, one would have to specify what system of coordination for pharmaceutical research could step in for (or improve on) the coordinating function that is at present taken over by the global market. Still, it seems to me that two major problems are already evident in the outline of the proposal. First, it is highly unlikely that the proposal has a chance of being actually implemented. The obstacles to implementation appear to be particularly great since only the abolishment of patent protection in all of the major markets would effectively remove the economic basis for private research. Second, even if it is taken for granted that the proposal addresses a matter of principle rather than one of options for actions, it would be unlikely to remove strong and partial interests from the health systems. Instead, the other actors of the health systems—in particular health insurers, medical associations, public research institutions, and patient organizations—keep their stakes. In Germany, public health insurers, for instance, have demonstrated repeatedly that they have rather specific interests concerning medical innovation. Since they compete with each other and with private health insurers for customers, they are strongly interested in controlling the costs of medical treatment. The medical associations, by contrast, tend to strongly oppose cost cuts. Still, their opposition is primarily directed against limitations on doctors' treatments and prescriptions, while they often seem to agree with health insurers as far as the pricing of pharmaceuticals is concerned. Other actors, such as the pharmacists' association, institutions of basic or clinical research, or patient organizations dedicated to specific diseases, bring their own and often diverging interests into play. For instance, the scientific reward system often bestows higher rewards for fundamental pharmaceutical achievements than for product development. It could therefore be feared that public pharmaceutical research would be biased in favor of academic rather than practical medical results. These observations illustrate that the interests of the pharmaceutical industry are not the only ones to have an impact on medical research. Whatever form socialized medicine and health research would take, it seems most unlikely to turn an area now dominated by strong and partial interests into one ruled by the impartial consideration of the most urgent needs.

Pluralism and the Advocate Model of Scientific Objectivity

The alternative to the exclusion of partial interests from pharmaceutical research could consist of making productive use of them in such a way that impartiality emerges as an overall effect. Such an idea is defended in this volume by Martin Carrier. According to Carrier, partial interests of many different actors could compete for realization in such a way that they control each other and thus preclude the undue realization of any single agenda. This mechanism relies on the plurality of partial interests, and, by making use of them as motivating factors both for competition and mutual control, it implicitly sanctions them. In its justification, Carrier refers to what could be called an "advocate model" of scientific objectivity. The advocate model deals with a diversity of opinions rather than of interests. It rejects the Baconian demand that the starting point of scientific inquiry should be free from theoretical or other preconceptions. Accordingly, scientific objectivity does not presuppose that scientists are unprejudiced, but rather that existing preconceptions are controlled and criticized in scientific debate (Carrier, this volume; cf. Longino 2002, chap. 6). Scientists are not then obliged to be severely critical of their own hypotheses. Rather, they might advocate the ideas they have put forward and present the strongest possible case for them, while leaving criticism and the production of the counterevidence largely to other scientists (Woodward and Goodstein 1996, 484–85). According to the pluralist position that results from the transfer of this model to the management of diverging interests in commercialized research, the competition between partial interests is seen to spur their mutual control and, ultimately, their neutralization.

Many of the measures that have been proposed and in part implemented in reaction to biasing interests in pharmaceutical research can be understood as aiming to establish or strengthen pluralism. Some have argued, for instance, that while industrial contributions to pharmaceutical research and development are indispensable, there should be an increased proportion of publicly funded research that does not carry with it similar commercial motives (Resnik 2000, 278). Public research institutes such as university hospitals and regulatory authorities (such as the U.S. Food and Drug Administration, or FDA) do, in fact, conduct such research.[5] In addition, biomedical research is also regularly funded by private trusts or patients' organizations. International organizations, in particular the World Health Organization (WHO), are also very active in co-ordinating and sometimes funding medical research (for example Oudshoorn 1998; WHO 2003). In Germany, the Institute for Quality and Cost Efficiency in the Health System, supported by the medical associations and public health insurers, has the task of evaluating medical therapies independent of industrial interests (IQWIG 2006). The position of the regulatory authorities in Europe has finally been strengthened by a central registry for all clinical studies. The aim is to improve the authorities' evaluation of industry-presented evidence by providing an overview of all studies that were planned or conducted by the company or other organizations. In this case, however, pluralism remains rather restricted as long as only the regulatory authorities, and not academic or clinical researchers or the public, have access to the registered data (Albrecht 2004; Antes et al. 2005).

Medical journals occupy a key position in clinical research and practice, since they are the main organs for the communication of clinical results. The publication of a study in a well-known and widely distributed medical journal not only earns the authors considerable scientific prestige but can also have a sustained influence on medical practice. Positive evaluations of drugs in these journals are therefore economically highly valuable for pharmaceutical companies. The medical journals have repeatedly used this position to try to implement better control of commercial interests. For instance, the editors of leading journals have decided that authors should declare conflicting interests. The sponsor of a study has to be named and authors' potential conflicts of in-terest due to remunerations or shareholdings must be made public (Davidoff et al. 2001). In addition, studies will only be published if they have been registered in a public directory before patient enrollment (DeAngelis et al. 2004).

The journals themselves are not free from commercial dependencies. There are reports, for instance, that journals have rejected papers because of fears that they might upset important advertisers (Dyer 2004; Bartens 2006). The most valuable asset of a scientific journal, however, is its credibility. Publications that are biased due to commercial interests thus threaten their own economic basis. From the viewpoint of pluralism, the actions taken by the medical journals can therefore be interpreted as attempts not only to retain their reputation as a source of trustworthy and reliable scientific information but also to protect the economic basis of their activities.

Scientific investigations of research and publication bias have increased greatly in recent years and have given rise to a new research program that has been called the "epidemiology of industry-sponsored research" (Montaner, O'Shaughnessy, and Schechter 2001). Such research is a powerful tool not only in determining how widespread biasing influences are, but in uncovering bias, it also criticizes it scientifically and thus helps to counter it by scientific means. Since according to pluralism, partial interests are to be tamed by criticism and control, scientific investigations can be taken to play a central role in the plural- ist management of interests. At the same time, however, it can only be in the interest of medical researchers themselves to retain or restore the reputation of medical science as a provider of objective knowledge, and doing so by scientific means could demonstrate the ability of the profession to defend its respect- ability in an autonomous way.

Finally, the pharmaceutical industry itself could profit from the control of biasing interests. The industry's worsening reputation not only might increasingly hamper attempts to carry through political aims but could also substantially weaken the effectiveness of promotional campaigns. In Germany, the prescription drug companies (including the German subsidiaries of inter- national companies) founded an association to control the marketing activities of its members. A code of conduct specifies the legitimate bounds of industry- physician relationships, for example concerning remunerations and gifts for the participation of physicians in industry-sponsored educational events or for talks given by clinical researchers. A court of arbitration was installed that includes a stage of appeal. According to this system of self-regulation, compa- nies that violate the code of conduct will be fined or publicly reprimanded (FS Arzneimittelindustrie 2006).

Conditions for the Pluralist Control of Commercial Interests

This overview indicates that the reactions to bias in pharmaceutical research and development come from a broad range of actors and lie on different levels. There are one-sided actions, such as the rules issued by the medical journals; state regulation, such as the legal order to public health insurers and medical associations to establish the Institute for Quality and Cost Efficiency in the Health System; and self-regulation, such as the pharmaceutical industry's mar- keting code and the rise of the scientific research program on research and pub- lication biases. The examples furthermore suggest that well-known national peculiarities in styles of regulation also play a role in how the problems that arise from the biasing interests are dealt with (cf. Daemmrich and Krücken 2000). The characteristic neocorporatist style of regulation shows up, for instance, in Germany, when it is regulated by law that actors of the health system have to establish the industry-independent institute for the evaluation of medical evidence, when the companies' "voluntary" mode of self-regulation preempts state regulation, or when access to the clinical trial registry is restricted to the regulation authorities. National peculiarities add to the many circumstances on which the eventual success of the measures depend, and which make it difficult to estimate in advance the prospects of success. Still, a number of general conditions for an effective pluralist control of partial interests seem plausible.

Greater transparency concerning special interests and available scientific evidence is the goal of a variety of measures, including journals' terms of publication and the European registry of clinical tests. Openness by itself is not yet a full remedy to bias, to be sure. The mere disclosure of an author's financial interest in the subject of a research paper does not suffice to restore the paper's reliability or credibility, for instance. Nor does knowledge of all clinical tests exclude the manipulation of a particular test. Nevertheless, openness concerning the interests and the full range of evidence allows all parties involved to better evaluate the presented results. If potentially biasing interests and the full range of evidence remained undisclosed, it would be hard for others to check whether interests influenced the results or whether they are presented selectively. Transparency is thus an important precondition for the mutual check of partial interests according to pluralism.

Still, for real competition among the partial interests to take place, more is needed than the disclosure of all interests and data. Participation in competition presupposes competences and resources that some of the interested parties might lack. The resources for research by clinical, academic, or regulatory organizations or by institutions of the public health systems therefore have to be sufficient to enable the parties to be active in the most important therapeutic areas and in cases in which they suspect serious bias of the existing results. Real competition does not presuppose that all parties have equal resources. Yet, from the perspective of pluralism, the resources of even the weakest party have to be sufficient for it to be able to enter the stage when its central interests are at stake.

However, even a competition based on sufficient and distributed resources and on transparency seems not to guarantee increased reliability and trustworthiness of medical knowledge. Admittedly, parties that ignore urgent medical needs for commercial reasons now have to expect that their studies will be criticized. Still, under the conditions of pluralism, criticism seems likely to come up not only if they violate scientific objectivity, but also if the results are contrary to the partial interests of some other party. Instead of giving in to criticism, whatever its nature and source, an agent motivated merely by partial interests could therefore also choose to increase her efforts to promote her standpoint. Similarly, the anticipated possibility of counterstudies does not necessarily deter an agent from biasing results. She could instead choose to falsify the results even more strongly. Perhaps a mean value between one's own and the opponent's results would seem most plausible to the wider audience, since the opponent is also known to pursue her own partial interests. Or a special audience may exist for which the sponsor's results will seem more trustworthy than those of the counterstudy, and it is this audience which is most important to the sponsor's interests. (Think, for instance, of patients who are critical with respect to orthodox, science-based medicine, and who have more confidence in natural cures, homeopathy, or traditional Chinese medicine.) In addition, since it will not be possible to check all steps or to reproduce all of the results, an organization that acts purely on selfish motives might try to bias results just in those cases in which control and replication are not to be feared.

If such scenarios seem unlikely to happen, it is not because the pluralist control of interests as such prevents them. Rather, it is because (and to the extent that) it seems unlikely that researchers or companies are so ruthless as to pursue their own interests so persistently. We would expect them to give in to justified criticism at an earlier or later stage, and stop trying to pursue their own interests up to the very end. But this means that the participants whom we imagine in the competition are not motivated merely by their own interests, but value to some degree the disinterestedness and objectivity of science as such.

Thus there are limits to the capacity of pluralism to secure the objectivity of research by competition and by mutual control of partial interests. While the advocate model of scientific rationality promises to manage prejudices effectively, the pluralist mechanism conceived according to this model seems much less potent in controlling competing interests. The crucial difference between pluralism and the advocate model, namely, that the one deals with diverging interests and the other with different opinions or scientific positions, seems to matter. From an abstract point of view, there seems to be a straightforward explanation for this difference. When research results are biased due to presuppositions or prejudices of the researchers, criticism of these results not only reveals that the results lack scientific justification, but at the same time can cause revision of the prejudices. As doxastic states, they are naturally the objects of scientific criticism, and as long as scientists are rational enough to revise presuppositions in the light of such criticism, the competition between differently prejudiced research programs seems likely to lead to objective overall results (Longino 2002, chap. 6). Diverging interests, by contrast, are much less susceptible to revision in the light of rational criticism. The insight that some of my interests are partial and influence the research results need not, by means of scientific rationality, lead to a revision of these interests. Neither does it demonstrate that the interests as such are unjustified. For example, companies are still justified to seek profits with their products despite the fact that these interests at times lead to unacceptable bias in the results of corporate research. While the influence of these interests on scientific research is problematic and can be subject to justified criticism, scientific rationality alone is not sufficient to alter the interests themselves as the source of this influence. Constant motivation not to let interests unduly influence research results is therefore required. In other words, motivation in favor of disinterestedness is necessary.

If disinterestedness retains its validity as a methodological and ethical standard also under the conditions of pluralism, genuinely moral mechanisms to enforce the norm can be put into place, for instance public affirmation of the norm, its formal inclusion in guidelines and codes, and disapproval of violations. Such measures with a definite moral content have also been implemented or are discussed with respect to pharmaceutical research. For instance, one of the aims of the above-mentioned system of self-regulation of the pharmaceutical industry is to keep companies from being pressured by economic competition to pursue illegal or immoral marketing strategies. The code thus aims to keep economic competition fair, and it was accordingly accepted by the competition authorities as a rule of competition. The moral character of the system of self-regulation is evidenced further by the fact that public rebuke is the maximum penalty. In scientific as in general public debate, the revelation of biasing influences of commercial interests regularly evokes considerable indignation. In medical journals, the publication of studies that show such influence is often accompanied by editorials that criticize the pharmaceutical industry and at times by scathing satire (e.g., Sackett and Oxman 2003). Newspapers and magazines regularly report on the results of these studies and amplify popular indignation (e.g., von Lutterotti 2003; Paulus 2004). Like mutual control of interests, the threat of public outrage can certainly reduce the proportion of genuinely moral motivation that the agents need in order to keep commercial and other self-directed interests under control. As claimed by Merton, the disinterestedness of science then does not seem to presuppose that scientists have particularly strong moral motives. The status of disinterestedness as a key element of the normative constitution of science is nevertheless confirmed by the expectation that individual moral motivation, the public enforcement of moral standards, and the pluralist control of biasing interests are together capable of maintaining the objectivity and impartiality of science even in the face of its progressive commercialization. Of course, this does not show that actual pharmaceutical research will fulfill these moral demands.

NOTES

1. For a detailed proposal of how interests should ideally influence the determination both of research priorities and the application of scientific findings, see Kitcher (2001, chap. 10).

2. See for example Strevens (2003) for defense of the claim that the priority rule, according to which reward for scientific achievements is given almost exclusively to the first producer of the new result, not only leads to an efficient allocation of resources, but is also fair since it matches reward to the contribution to society.

3. Between 1981 and 1990, 92 percent of the chemically new drugs approved by the U.S. Food and Drug Administration were developed by the pharmaceutical industry (Kaitin, Bryant, and Lasagna 1993).

4. Moral restrictions on corporate research might apply, of course, if the research not only serves the satisfaction of the needs of the company's customers, but at the same time harms (other) people. Corporate research on the development of weapons, for example, is certainly subject to such additional restrictions, for example, concerning the kinds of weapons to be developed and their expected proliferation. However, to the extent that pharmaceutical development is intended to benefit some individuals without harming others, such restrictions need not apply here.

5. The FDA also seems not to be free from industry influence, as the Vioxx case brought to light. With the introduction of an industry user fee for new drug approvals in 1992, the share of the budget that was spent by the FDA's Center for Drug Evaluation for the approval of new drugs rose from 53 percent in 1992 to 79 percent in 2003, with a corresponding decline of the spending share on research concerning the safety of approved drugs. See Harris (2004). (My thanks to Justin Biddle for drawing my attention to these interwoven interests.)

REFERENCES

Albrecht, Harro. 2004. "Blockiertes Register." *Die Zeit* 18 (22 April 2004).

Antes, Gerd, et al. 2005. "Plädoyer für die Einrichtung eines öffentlichen Registers." *Deutsches Ärzteblatt* 102 (27), 8 July 2005: 1937.

Als-Nielsen, Bodil, et al. 2003. "Association of Funding and Conclusions in Randomized Drug Trials." *Journal of the American Medical Association* 290:921–28.

Bartens, Werner. 2006. "Eiertanz der Gutachter: Wie ein medizinischer Fachverlag dem Druck der Pharmaindustrie nachgab und eine kritische Artikelserie stoppte." *Süddeutsche Zeitung*, 19 September 2006.

Bayles, Michael D. 1989. *Professional Ethics*. 2nd ed. Belmont: Wadsworth.

Brown, James R. 2002. "Funding, Objectivity and the Socialization of Medical Research." *Science and Engineering Ethics* 8:295–308.

Collier, Joe, and Ike Iheanacho. 2002. "The Pharmaceutical Industry as an Informant." *Lancet* 360:1405–9.

Daemmrich, Arthur, and Georg Krücken. 2000. "Risk versus Risk: Decision-making Dilemmas of Drug Regulation in the United States and Germany." *Science as Culture* 9:505–34.

Davidoff, Frank, et al. 2001. "Sponsorship, Authorship, and Accountability." *Lancet* 358:854–56.

Davidson, Richard A. 1986. "Source of Funding and Outcome of Clinical Trials." *Journal of General Internal Medicine* 1:155–58.

DeAngelis, Catherine D., et al. 2004. "Clinical Trial Registration." *Journal of the American Medical Association* 292:1359–62.

Djulbegovic, Benjamin, et al. 2000. "The Uncertainty Principle and Industry-Sponsored Research." *Lancet* 356:635–38.

Dyer, Owen. 2004. "Journal Rejects Article after Objections from Marketing Department." *British Medical Journal* 328:244.

Etzkowitz, Henry. 1997. "The Entrepreneurial University and the Emergence of Democratic Corporatism." In *Universities and the Global Knowledge Economy: A Triple Helix of University-Industry-Government Relations*, ed. Henry Etzkowitz and Loet Leydesdorff, 141–52. London: Pinter.

Food and Drug Administration. 2005. "Label Alinia (nitazoxanide)," version as of 16 June 2005. http://www.fda.gov (last accessed 3 November 2006).

Friedberg, Mark, et al. 1999. "Evaluation of Conflict of Interest in Economic Analyses of New Drugs Used in Oncology." *Journal of the American Medical Association* 282:1453–57.

FS Arzneimittelindustrie. 2006. *Webpages Freiwillige Selbstkontrolle Arzneimittelindustrie*. http://www.fs-arzneimittelindustrie.de (last accessed 3 November 2006).

Gibbons, Michael E., et al. 1994. *The New Production of Knowledge: The Dynamics of Science and Research in Contemporary Societies*. London: Sage.

Grice, Paul. 1989. *Studies in the Way of Words*. Cambridge, MA: Harvard University Press.

Harris, Gardiner. 2004. "At F.D.A., Strong Drug Ties and Less Monitoring." *New York Times*, 6 December 2004.

Hirsch, Laurence J. 2002. "Conflicts of Interest in Drug Development: The Practices of Merck & Co., Inc." *Science and Engineering Ethics* 8:429–42.

IQWIG. 2006. *Webpages Institut für Qualität und Wirtschaftlichkeit im Gesundheitswesen.* http://www.iqwig.de (last accessed 3 November 2006).

Kaiser, Thomas, et al. 2004. "Sind die Aussagen medizinischer Werbeprospekte korrekt?" *Arznei-Telegramm* 35 (2): 21–23.

Kaitin, Kenneth I., N. R. Bryant, and Louis Lasagna. 1993. "The Role of the Research-Based Pharmaceutical Industry in Medical Progress in the United States." *Journal for Clinical Pharmacology* 33:412–17.

Kitcher, Philip. 2001. *Science, Truth, and Democracy.* Oxford: Oxford University Press.

Longino, Helen. 2002. *The Fate of Knowledge.* Princeton: Princeton University Press.

Lutterotti, Nicola von. 2003. "Das Schweigen der Forscher: Veröffentlicht wird oft nur das, was gefällt." *Frankfurter Allgemeine Zeitung* 293, 17 December 2003.

Melander, Hans, et al. 2003. "Evidence B(i)ased Medicine—Selective Reporting from Studies Sponsored by Pharmaceutical Industry: Review of Studies in New Drug Applications." *British Medical Journal* 326:1171–75.

Merton, Robert K. [1942] 1973. "The Normative Structure of Science." In *The Sociology of Science*, 267–78. Chicago: University of Chicago Press.

Montaner, Julio S. G., Michael V. O'Shaughnessy, and Martin T. Schechter. 2001. "Industry Sponsored Clinical Research: A Double-Edged Sword." *Lancet* 358:1893–95.

Oudshoorn, Nelly. 1998. "Shifting Boundaries between Industry and Science: The Role of the WHO in Contraceptive R&D." In *The Invisible Industrialist: Manufactures and the Production of Scientific Knowledge*, ed. Jean-Paul Gaudillière and Ilana Löwy, 345–79. London: Macmillan.

Paulus, Jochen. 2004. "Die Tricks der Pillendreher: Wie Pharmafirmen mogeln, damit Studien die gewünschten Resultate zeigen." *Die Zeit* 18 (22 April 2004).

Ramsay, Sarah. 2001. "No Closure in Sight for the 10/90 Health-Research Gap." *Lancet* 358:1348.

Resnik, David B. 1998. *The Ethics of Science: An Introduction.* London: Routledge.

———. 2000. "Financial Interests and Research Bias." *Perspectives on Science* 8:255–85.

Romark. 2006. *Webpages Romark Laboratories.* http://www.romarklabs.com (last accessed 3 November 2006).

Rossignol, Jean-François, and Raymond Cavier. 1975. "2-benzamino-5-nitrothiazoles." *Chemical Abstracts* 83:28216n.

Sackett, David L., and Andrew D. Oxman. 2003. "HARLOT plc: An Amalgamation of the World's Two Oldest Professions." *British Medical Journal* 327:1442–45.

Shrader-Frechette, Kristin. 1994. *Ethics of Scientific Research.* Boston: Rowman and Littlefield.

Strevens, Michael. 2003. "The Role of the Priority Rule in Science." *Journal of Philosophy* 100:55–79.

White, A. Clinton, Jr. 2003. "Nitazoxanide: An Important Advance in Anti-Parasitic Therapy." *American Journal of Tropical Medicine and Hygiene* 68:382–83.

World Health Organization (WHO). 2002. *Guidelines for Drinking-water Quality —Addendum: Microbial Agents in Drinking Water.* 2nd ed. Geneva: World Health Organization.

————. 2003. "US$30 Million Research Effort to Develop New Test for Deadly Infectious Diseases." *World Health Organization Press Release*, 22 May 2003. http://www.who.int/mediacentre/news/releases/2003/en/ (last accessed 3 November 2006).

Woodward, James, and David Goodstein. 1996. "Conduct, Misconduct and the Structure of Science." *American Scientist* 84:479–90.

Ziman, John. 2003. "Non-instrumental Roles of Science." *Science and Engineering Ethics* 9:17–27.

INDEX

accuracy: as cognitive virtue, 4, 45; and expertise, 155; and precision (*see* precision); within science, 9, 57, 152; within theories, 4

Adler, Alfred, 150

androcentrism, 70, 88–89, 94

appraisal, theory, 3, 5, 78, 81, 96, 222

Aristotle, 117, 157

artificial pairs, 26, 39–40. *See also* observationally equivalent theories

Bacon, Francis, 12, 87, 153, 157, 213

Baconianism: regarding scientific goals, 112–15, 223, 245

Bayh-Dole Act, 191, 214n3. *See also* research

Bayesianism, 32. *See also* induction: probabilistic

biochemistry, 71, 147, 217; and in-vivo/in-vitro research 150, 152

biometrics. *See* statistics: biometrics

Bloor, David, 40n2, 206

Bohm, David. *See* quantum mechanics: hidden variable interpretation of

BSE (Mad Cow Disease), 131

causality: causally effective, 70; causal explanation, 56; causal interactions, 74; causal processes, 220; causal relations, 29, 72; contextualized causal relations, 220, 224; unicausal models, 70, 74

chemistry, 89, 217, 224

clinical trials, 199, 204, 207, 209, 248; corruption in, 192–96

cognitive value, 2–5, 7–8, 45–47, 69, 72–78, 81, 217; constitutive of science, 9, 45, 79, 113–15, 117, 220; and feminism, 72–78; as heuristics, 115; and methodology, 211–12

Collins, Harry, and Evans, Robert, 160–79, 180n3, 180n6, 180n8, 181n14, 181n17, 182n23

commercialization of science. *See* science: commercialization of. *See also* research: commercialization of

complexity of interaction. *See* cognitive value

communalism. *See* communism

communism, 189, 214, 219

competition: between theories, 14; economic, 11, 213, 250; as incentive for research, 225–26, 245, 248–50

confirmation, theory, 7, 20–21, 26–32, 40; and Duhem's holism 23–24; hypothetico-deductive account of, 26–28, 31, 41; relevance to evidential support, 28–29. *See also* induction; justification: inductive

conservatism, 2–3, 70, 115

consistency, theoretical, 3, 69, 70, 72–73

constitutive value. *See* cognitive value

contextual value, 45, 47, 52, 64

controversy: ambivalence in science toward, 167–70

cultured pairs, 25–26, 36–39. *See also* observationally equivalent theories